Towards Understanding Receptors

Towards Understanding Receptors

Current Reviews in Biomedicine 1

Edited by John W. Lamble

Foreword by G. Alan Robison

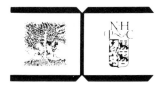

1981

ELSEVIER/NORTH-HOLLAND BIOMEDICAL PRESS
Amsterdam – New York – Oxford

ISBN: 0–444–80339–4

Published by
Elsevier/North-Holland Biomedical Press BV
P.O. Box 1527
1000 BM Amsterdam
The Netherlands

Sole distributors worldwide except for the U.S.A. and Canada
Elsevier/North-Holland Biomedical Press
68 Hills Road
Cambridge CB2 1LA
United Kingdom

Sole distributors for the U.S.A. and Canada
Elsevier/North-Holland Inc.
52 Vanderbilt Avenue
New York, NY 10017, U.S.A.

PRINTED IN THE UNITED KINGDOM

Foreword

Our understanding of what receptors are and how they work has increased spectacularly during the past 5 years. It was not so long ago that it was possible for informed investigators to speculate that the demise of the receptor concept was close at hand, since, according to that view, we would soon know what receptors were, following which we would be able to dispense with the concept. We now know better. Receptors are recognized as real rather than conceptual entities, as important for survival as any other body constituent, and the increasingly successful attempts to understand them in molecular terms has become one of the most exciting and rapidly moving areas in all of molecular biology.

Following the introduction of the receptor concept by Ehrlich and Langley, as summarized in this volume by Parascandola, progress towards understanding receptors was initially very slow. Clark summarized the progress that had been made to that point in 1937, in his now classic monograph on *General Pharmacology*[1]. The purpose of general pharmacology, as Clark defined it, was to understand drug action in chemical terms, and most of the major questions with which we are now grappling were posed by Clark in a remarkably prescient way. Nevertheless, while this book had great influence on a number of individuals, including notably the late Earl Sutherland, progress continued to be slow for many additional years. It is interesting and may even be instructive to speculate on this slow progress, and I would offer the following possible reasons for it.

One reason, to which I have already alluded, was the tendency of many pharmacologists, both before and after Clark, to regard receptors as conceptual entities only, to suppose that what we referred to as receptors for the sake of convenience would in reality turn out to be something else. This was damaging because it was an empty doctrine that could not be attacked by way of the scientific method, and further damaging because there was nothing in it to stimulate the interest or excitement of young investigators. The geneticists and immunologists of the day did not make the same mistake or at least not to the same extent, and I would suggest that it is primarily for that reason that the chemical nature of genes and antibodies began to be uncovered before that of receptors. There is, after all, no point in trying to understand something if one is convinced that it does not exist.

A related reason, which actually contributed to the first, was the failure to distinguish between two major categories of drugs. These are the drugs that occur naturally within the body (hormones, neurotransmitters, autacoids, and the like), on the one hand, and the xenobiotics which are foreign to the body, on the other. The significance of this distinction is that receptors have evolved to interact with and mediate the effects of the naturally occurring drugs. Xenobiotics, by contrast, may interact with any of a large variety of cellular and extracellular constituents to produce their effects. This was appreciated from an early date, and was emphasized by Clark. The mistake was in subsequently defining whatever a drug interacted with as the 'receptor' for that drug. One of the first drugs whose mechanism of action was understood was carbon monoxide, which was found to interact with hemoglobin, thereby interfering with hemoglobin's ability to carry oxygen. By extrapolating from this and other examples, it seemed reasonable to suppose that when other receptors for other drugs were understood, they too would turn out to be something else, and that the substances imagined by Langley and Ehrlich would eventually be defined out of existence.

It is of course clear that some xenobiotics may interact with receptors, but it is interesting to realize that while all of the naturally occurring drugs that are known act as agonists, most of the xenobiotics that interact with receptors do so as antagonists. The exceptions to this rule have contributed disproportionately to the terminology we still use, as in the case of muscarinic and nicotinic receptors. The question of *why* agonists are agonists, or why antagonists are not, might properly be regarded as the central question of molecular pharmacology. More of this later.

Another related reason for the slow progress in this field was the reluctance on the part of many pharmacologists to utilize the scientific method, or what Platt[2] has referred to as the method of strong inference. One of the most striking features of molecular biology, since it has come to be thought of as a separate discipline, has been the systematic utilization of the scientific method. It was applied early in genetics and immunology, later in pharmacology, and it is not surprising that the recent rapid progress in our understanding of receptors has coincided to a remarkable degree with the more widespread use of this method.

But as a final reason for the initially slow progress in this field I might suggest, what may or may not seem obvious to future historians, that the subject has been very difficult. It is possible to look back at the state of biology at the beginning of this century, and at the work of Ehrlich and Langley in particular, to see how unlikely it would have seemed that an understanding of how hereditary information was transmitted, or of how the immune system functioned, would have preceded an understanding of what receptors are and how they work. The receptor problem would surely have seemed simpler, and yet it is also possible to see, in retrospect, that this apparent simplicity might have been misleading. I am thinking here primarily of the rapidity of drug action. It has now been established, for example, that the initial rate of ion translocation in response to acetylcholine (in membrane vesicles prepared from the electric organ of the eel, measured at low temperatures) is of the order of 6000 ions per *millisecond* per receptor-formed channel[3], which can be contrasted with the scale of hours or even days on which genetic and immunological events can be measured. Techniques for measuring such fast reactions, which may be slow compared to the rate at which receptors are activated in the first place (corresponding in this example to the rate at which the channels are formed), were not available until recently. Whether this defense of pharmacology's slow rate of progress during the first two thirds of this century will hold up remains to be seen. It is nevertheless a fact that it was not possible to study receptor-mediated events in a cell-free system, even under steady state conditions, until the discovery of cyclic AMP by Sutherland and his colleagues in 1957. Although many other antecedents could be pointed to (and are pointed to in the pages to follow), I might suggest that the most important for the progress that is currently being made in this field was the demonstration by Sutherland and Rall that adenylyl cyclase in the particulate fraction of a dog liver homogenate could be stimulated by glucagon and epinephrine. Even after that, progress was slow for many years, but perhaps not much slower than progress in molecular genetics following Avery's demonstration of the transforming activity of DNA.

But enough of the past. Surely it is the present that excites us, and the present, insofar as receptors are concerned, is very exciting indeed. Progress in this area is being made at an incredible rate, to the point where it is hardly possible to keep track of the number of books that are being published in it, let alone the new research papers that are appearing on a weekly basis. For those not already immersed in the subject, a highly readable over-view is badly needed, and this is precisely what is provided by this collection of review articles from *Trends in Pharmacological Sciences.*

The inaugural issue of *TIPS* appeared in April 1979, by which time the current revolu-tion in molecular pharmacology was well underway, and it was an ideal time to begin chronicling the progress that was and is occurring. All of the reviews in *TIPS* are quite short, partly to enable a good variety of topics to be accommodated in each issue, but also to maximize the impact of each article on the reader. The number of references per article has been deliberately kept to a minimum, but is sufficient to direct the reader to more detailed accounts. This volume is a collection from the first 2 years of *TIPS* of reviews dealing in one way or another with the question of what receptors are and how they func-tion.

A point that might go without saying is that in any field as rapidly moving as this one, reviews are generally out of date within a week or so of the day they are published. This is as true of articles that may have taken the authors months to write, and that may contain many hundreds of references, as it is of the shorter articles published in *TIPS*. And yet all good reviews are in another sense timeless, and will never go out of date. Perhaps the epitome of this would be Clark's review mentioned previously. New advances are con-tinuously being made, but a good review will bring together the new facts that have been established to that point, together with the new insights that have been developed from them, with the result that the past can be better understood and future progress better appreciated. They are like the cogs in a ratchet machine, taking us to a higher level of understanding with each turn of the wheel. I would like to submit that this volume con-tains a higher proportion of such timely reviews than most of the collections I have seen.

The reviews are arranged not in the sequence in which they were published in *TIPS*, but rather in groups according to the type of receptor or group of receptors under discussion. The collection begins, following Parascandola's historical article, with three papers deal-ing with receptors in general, and I would especially like to recommend Trevor Franklin's fine article on binding energy and receptor activation. Franklin approaches the central question of molecular pharmacology by asking why potent agonists seem to have less affinity for their receptors than partial agonists or antagonists, which seems analogous to the enzymological question of why good substrates (with high V_{max}) tend to have lower apparent affinities than do poor substrates or competitive inhibitors. Enzymology developed much earlier as a branch of molecular biology than did the study of receptors, and concepts of enzyme action have frequently been helpful in efforts to understand receptor function. This continues to be the case today, but Franklin's article reminds us how rapidly the gap between these two areas is being reduced. It is even possible to imagine that future advances in our understanding of receptors may contribute to a better understanding of how enzymes work, instead of the other way around, as has always been the case in the past. The reason agonists have greater intrinsic activity or efficacy than par-tial agonists or antagonists remains to be established, but Franklin asks the question in a more meaningful way than before, and points to the kind of approach that will probably have to be taken to achieve the answer.

Colquhoun's article is also extremely good. It analyses in an interesting and helpful way the speed of drug action, which I referred to earlier as one of the major impediments to progress in this field. The velocity cited previously was actually drawn from a more recent paper to which readers of Colquhoun's article are referred[3].

Triggle's article deals with the important and increasingly well understood phenomenon of tachyphylaxis or desensitization. I previously compared progress in immunology and pharmacology, and it is interesting to realize how intertwined this progress was at one time. Ehrlich's interest in both subjects is well known (see also Parascandola on this), but another dualist was Charles Richet, who received the Nobel Prize in 1913. His discovery of anaphylaxis was made as a serendipitous side-result of his interest in the pharmacology of marine toxins, and he subsequently coined the term tachyphylaxis in the mistaken belief that it too was an immunological phenomenon. Whether pharmacology and immunology will seem to converge more than they diverge as we learn more (see also the review by Foreman on this point) remains to be seen. In the meantime, it is now clear that an important mechanism of tachyphylaxis involves a reduction in the activatability and subsequently the number of receptors[4], and Triggle's article provides a good introduction to this field of study. The psychopharmacological significance of tachyphylaxis is emphasized later in this volume in the article by Sulser. This important article nicely illustrates how a better understanding of receptors may lead to a better understanding of the mechanism of action of clinically useful xenobiotics. That may in turn enable us to use these drugs more wisely and effectively.

Much of the recent progress in molecular pharmacology has stemmed from studies of the relation between adrenergic receptors and adenylyl cyclase, and the article by Strosberg and his colleagues, together with the first of the two articles by Lefkowitz and Hoffman, provides an excellent introduction to this very active field of research. The article by Strosberg *et al.* is incidentally the only one in this volume that was published initially not in *TIPS* but in one of its sister publications, *Trends in Biochemical Sciences*. This raises an important and frightening question of territoriality. As we have known since 1937, all attempts to understand drug action in chemical terms are by definition part of the science of *general pharmacology*. But might not biochemical studies of receptors also be construed as biochemistry? Fortunately I have been asked only to write a foreword, not to settle fundamental disputes of this nature, and I can only say how pleased I am that the article by Strosberg *et al.* has been included. It can be noted that the guanyl nucleotide binding protein to which this and some of the other articles in this volume refer has now been purified to homogeneity[5] and the kinetics of its activation by receptors studied[6]. The utility of enzymological concepts was mentioned previously, and I might suggest that a helpful way to think of this protein (variously referred to in the literature as N, G, or G/F) would be as a substrate for the active form of the receptor. The reaction catalysed would be the dissociation of GDP followed by the association with GTP followed by the release of the product. This way of looking at things makes a certain amount of sense, at least in my own mind.

The importance of calcium and the potential importance of changes in phospholipid turnover in drug action were recognized many years ago, and some of the more recent research connecting the two is reviewed in the articles by Berridge and by Michell and Kirk. Berridge incidentally points to an increasingly troublesome problem of nomenclature, related in a sense to the earlier tendency of pharmacologists to believe that receptors would eventually be defined out of existence. Most of the receptors we know about were named at a time when nothing was known about them beyond their ability to mediate the effects of this or that naturally occurring drug, and it did not seem to matter much what they were called (see also the articles by Ahlquist and by Kebabian and Cote on this point). But now that we are beginning to discern the similarities and differences between different groups of receptors in biochemical terms, it would seem desirable to develop terminology that would reflect this new knowledge. Perhaps the International Union of

Pharmacology could play a useful role here by setting up a Commission on Nomenclature analogous to the one established by the International Union of Biochemistry. Just as the IUPAC-IUB Commission was helpful to biochemists interested in communicating with each other about enzymes, I believe an analogous Commission sponsored by IUPHAR could facilitate communication between and among pharmacologists interested in receptors. In the meantime, we can only reiterate that there is no necessary relation between what receptors are called and what they do.

The point was made previously that a better understanding of receptors may lead to a better understanding of the mechanism of action of clinically useful xenobiotics. The converse of this, the study of xenobiotics leading not only to a better understanding but to the actual discovery of an important group of naturally occurring drugs, is illustrated in the articles on opiate receptors by Hughes and Simon.

Inspection of these and the other fine articles in this volume illustrates what may go without saying, that our knowledge of the chemical nature and function of some types of receptors is much greater than of others. But all of these articles contain useful information, and it can be confidently predicted that as our knowledge of the structure and function of one type of receptor increases, the resulting insights will be applied ever more rapidly to understanding others.

This begs the interesting question of how general the principles of pharmacology will prove to be. What features will all agonists have in common and what features will all receptors have in common, when our understanding of their interactions is more complete? It is yet too early to answer this question, although that should surely be one of our goals, and I believe Franklin's article, mentioned previously, suggests the outline of at least one common feature. It may be helpful in the meantime to consider the similarities and differences between receptors located in membranes, on the one hand, and the receptors for steroid hormones, on the other. Progress in understanding these two groups of receptors seems in general to have moved along parallel lines, with an inexplicable minimum of deliberately shared insights. Most of the articles in this volume are concerned with membrane receptors, reflecting the greater attention which classical pharmacologists (as opposed to endocrinologists) paid to these receptors in the past. It is for this reason that the articles in this volume on steroid receptors, by Raynaud and Walters and their colleagues, are to be especially treasured. A number of significant insights, not to mention some interesting problems with which many pharmacologists are unfamiliar, will be gleaned from reading these articles.

This volume concludes with three articles not about receptors for naturally occurring drugs, as we traditionally think of them, but about receptors for other substances which may serve as models for drug action. These are the articles by Goodford on hemoglobin, by Foreman on IgE receptors (which may be quite analogous to insulin receptors), and by Crosby and DuBois on taste receptors. All of these articles are of interest, quite apart from the light they may shed on the function of other types of receptors.

Mentioning insulin receptors, which are not discussed in this volume, reminds me to point out that this volume was not expected to be completely comprehensive although in this respect it rivals most others published in the field. It is quite simply, as I said before, a collection of articles from the first 2 years of *TIPS'* existence and future volumes will repair any omissions. It is a superb collection which I can recommend enthusiastically to expert and novice alike. Those who regard themselves as experts in one of the areas covered in this book are after all likely to be novices in another. But it may be especially valuable to graduate students coming to the subject for the first time, both those who are trying to brush up on their own and those who are taking a course in the subject. It could even be recommended as a text for such a course, for I know of no textbook that offers a more readable introduction to the subject.

One of the things I like best about it is that so many of the authors have allowed their interest and excitement to show through. With a few exceptions, such as Clark's great monograph and the first edition of Goodman and Gilman, most books about drug action have not been fun to read. I think most readers will agree with me that this is one of those exceptions, and I would like to thank the authors and editor for making it possible.

G. ALAN ROBISON

Department of Pharmacology,
University of Texas Medical School,
Houston, Texas, U.S.A.

References

1 Clark, A. J. (1937) *General Pharmacology: Heffter's Handbuch der Experimentellen Pharmakologie* (Heubner, W. and Schuller, J., eds), Vol. 4, Springer-Verlag, Berlin. Reprinted 1971
2 Platt, J. R. (1964) *Science*, 146, 347–353
3 Hess, G. O., Aoshima, H., Cash, D. J. and Lenchitz, B. (1981) *Proc. Natl Acad. Sci. U.S.A.* 78, 1361–1365
4 Su, Y. F., Harden, T. K. and Perkins, J. P. (1980) *J. Biol. Chem.* 255, 7410–7419
5 Northrup, J. K., Sternweis, P. C., Smigel, J. D., Schleifer, L. S., Ross, E. M. and Gilman, A. G. (1980) *Proc. Natl Acad. Sci. U.S.A.* 77, 6516–6520
6 Citri, Y. and Schramm, M. (1980) *Nature (London)*, 287, 297–300

Preface

This volume is the first in a series to be produced on receptor research and other biomedical topics. It consists largely of reviews from *Trends in Pharmacological Sciences* although one paper is from the sister publication *Trends in Biochemical Sciences*. Future volumes will perpetuate this cross-disciplinary approach.

The difficulties of establishing a basis of knowledge and then maintaining it in a burgeoning scientific speciality are well known. Receptor research is important for all students and practitioners in the biomedical field and the current volume makes readily accessible at a modest price much essential information. Above all, the material in this book is *readable* and I would like to record my appreciation of the great efforts the authors have made to communicate their specialized knowledge to those not directly involved in the field.

I must also express my deep gratitude to Dr G. Alan Robison for his advice during the assembly of this book and for his contribution of an excellent foreword.

JOHN W. LAMBLE

Contents

Foreword, *G. Alan Robison*   v

Editor's Preface   xi

Origins of the receptor theory, *John Parascandola*   1

Binding energy and the excitation of hormone receptors, *T. J. Franklin*   8

How fast do drugs work?, *D. Colquhoun*   16

Desensitization, *D. J. Triggle*   28

The acetylcholine receptor, *Jérôme Giraudat and Jean-Pierre Changeux*   34

Early work on the adrenergic mediator: how to go wrong, *Z. M. Bacq*   44

Adrenoceptors, *Raymond P. Ahlquist*   49

Towards the chemical and functional characterization of the β-adrenergic receptor, *A. Donny Strosberg, Georges Vauquelin, Odile Durieu-Trautmann, Colette Delavier-Klutchko, Serge Bottari and Claudine Andre*   53

New directions in adrenergic receptor research, Part I, *Robert J. Lefkowitz and Brian B. Hoffman*   61

Identification and significance of beta-adrenoceptor subtypes, *Stefan R. Nahorski*   71

β-Adrenergic receptor subtypes, *G. Leclerc, B. Rouot, J. Velly and J. Schwartz*   78

New directions in adrenergic receptor research, Part II, *Robert J. Lefkowitz and Brian B. Hoffman*   84

Actions of hormones and neurotransmitters at the plasma membrane: inhibition of adenylate cyclase, *Karl H. Jakobs and Günter Schultz*   90

Presynaptic receptors and the control of noradrenaline release, *Klaus Starke*   94

New perspectives on the mode of action of antidepressant drugs, *F. Sulser*   99

Dopamine receptor: from an *in vivo* concept towards a molecular characterization, *Pierre Laduron*   105

Dopamine receptors and cyclic AMP: a decade of progress, *John W. Kebabian and Thomas E. Cote*   112

Clinical relevance of dopamine receptor classification, *Donald B. Calne*   118

Receptors and calcium signalling, *Michael J. Berridge*   122

Why is phosphatidylinositol degraded in response to stimulation of certain receptors?, *Robert H. Michell and Christopher J. Kirk*   132

Organization and transduction of peptide information, *R. Schwyzer*   139

Angiotensin receptors and angiotensinase activity in vascular tissue, *Djuro Palaic*   147

Peripheral opiate receptor mechanisms, *John Hughes*   151

Opiate receptors: some recent developments, *E. J. Simon*   159

Histamine H₁ receptors in the CNS, *J. M. Young*   166

GABA receptors, *S. J. Enna*   171

The role of gamma-aminobutyric acid in the action of 1,4-benzodiazepines,
    E. Costa   176

Searching for endogenous benzodiazepine receptor ligands, *Claus Braestrup and
    Mogens Nielsen*   184

The use of interaction kinetics to distinguish potential antagonists from agonists,
    J. P. Raynaud, M. M. Bouton and T. Ojasoo   192

Ubiquitous effects of the vitamin D endocrine system, *Marian R. Walters,
    Willi Hunziker and Anthony W. Norman*   200

The haemoglobin molecule: is it a useful model for a drug receptor?,
    Peter Goodford   205

Receptor-secretion coupling in mast cells, *John Foreman*   217

Sweeteners and receptor sites, *Guy A. Crosby and Grant E. DuBois*   223

Origins of the receptor theory

John Parascandola

School of Pharmacy, University of Wisconsin, Madison, Wisconsin 53706, U.S.A.

In the inaugural issue of 'Trends in Pharmacological Sciences', A. W. Cuthbert described the last two decades in pharmacology as "the age of the receptor"[1] and E. J. Ariëns noted that for decades the receptor concept has been "indispensable for discussing and understanding the mode of action of pharmaca"[2]. Textbooks and review articles often call attention to the fact that the receptor theory dates back to the work of Paul Ehrlich (1854–1915)[3] and John Newport Langley (1852–1925)[4,5,6] in the early twentieth century, but generally do not provide much more information than that about the origins of the concept. The purpose of this article is to examine the development of the receptor concept by Ehrlich and Langley.

Early speculations

As an interest developed in the relationship between chemical structure and pharmacological activity in the second half of the nineteenth century, it is not surprising that pharmacologists such as Thomas Fraser and Thomas Lauder Brunton speculated that the physiological action of drugs was due to a chemical reaction between the drug and some constituent of the cell. Some pharmacological evidence for this view was provided by the experimental work of J. N. Langley on pilocarpine, begun in 1874 while he was still a student of physiology at Cambridge University, England. In a paper published in 1878, he explained the observed antagonism between atropine and pilocarpine by assuming that these drugs compete for a specific substance in the body. He commented: ". . . we may, I think, without much rashness, assume that there is a substance or substances in the nerve endings or gland cells with which both atropin and pilocarpin are capable of forming compounds. On this assumption then the atropin or pilocarpin compounds are formed according to some law of which their relative mass and chemical affinity for the substance are factors".

This statement contains the germ of the receptor theory, but Langley did not follow it up for another quarter of a century.

The side chain theory of Paul Ehrlich

In the same year that Langley published the above-mentioned paper, Paul Ehrlich received his M.D. from the University of Leipzig. His thesis dealt with histological staining, and his interest in dyes was to influence much of his later work. For his 'Habilitationsschift' at the University of Berlin (1885), for example, Ehrlich chose to investigate the ability of different tissues to reduce certain dyestuffs. This study is relevant to our present concern because it contains the first indication of his famous side chain theory. He adopted Edward Pflüger's view that protoplasm can be envisioned as a giant molecule consisting of a chemical nucleus of special structure which is responsible for the specific functions of a particular cell (e.g. a liver cell) with attached chemical side chains. The latter, according to Ehrlich, were more involved in the vital processes common to all cells, such as nutrition and oxidation.

It was this side chain concept that Ehrlich used in 1897 to explain the neutralization of bacterial toxins by antibodies. His immunological experiments convinced him that the interaction between a toxin and an antitoxin was chemical in nature and stoichiometric. It must be assumed, he argued, that the ability to combine with the antitoxin is attributable to the presence in the toxin of a particular group of atoms with a specific affinity for another group in the antitoxin.

How is it, Ehrlich asked, that there are antibodies in the blood which can specifically combine with toxin molecules? He found it hard to believe that the cell could manufacture as needed antibodies specifically tailored to combine with various kinds of toxins. Ehrlich assumed, therefore, that these must exist beforehand, and must normally serve other physiological functions. He suggested that perhaps one type of side chain carries an atom group with a specific combining property for an atom group in a particular toxin, such as diphtheria toxin. This side chain is normally involved in ordinary physiological functions, and it is merely coincidental that it can combine with a toxin. Combination with the toxin, however, renders the side chain incapable of performing its normal physiological function. The cell then produces more of

Paul Ehrlich. (Courtesy of National Library of Medicine)

this side chain to make up for the deficiency, but it overcompensates. Excess side chains are produced, and these break away from the cell and are released into the bloodstream. These excess side chains in the blood are what we call antibodies or antitoxins, and they neutralize the toxin molecules by combining with them.

Ehrlich sometimes referred to the side chains as 'receptors', a term he later came to use with respect to drugs as well as toxins. He also went on to distinguish between the toxophore group of a toxin, which gives it its toxic properties, and the haptophore group, which fixes the drug to the cell and allows the toxophore group to exert its action. This view might well have been derived from an analogy with dyes, where the chemical group that fixes the dye to the fabric is generally different from the one responsible for color.

Reluctance to apply the concept to drugs

Accounts of Ehrlich's work generally stress the relationship between his views on the binding of toxins and on the binding of drugs. It is often stated or implied that the latter followed directly from the former. Yet if we examine Ehrlich's work more closely, we find that it took him about ten years to apply his side chain theory to the problem of drug action. In fact, at first he specifically denied that drugs were anchored to the cell in the same way as toxins.

In an address delivered in 1898, for example, he stated that it is not likely that

John Newport Langley. (Courtesy of National Library of Medicine)

a firm combination is formed between a drug and the cell, such as in the case of the toxin–antitoxin situation. He pointed out that many drugs can easily be extracted from tissues by solvents, thus they cannot be firmly bound. In addition, the action of many drugs is of a transitory nature and readily reversible, again arguing against the formation of a firm chemical bond. In his Croonian Lecture in 1900, Ehrlich again stressed that drugs, as opposed to toxins, do not possess atom groups which enter into combination with corresponding groups in the protoplasm. Within the cell, they enter only into unstable combinations, notably with the "not really living parts" of the cell.

Yet Ehrlich was well aware of the fact that many drugs do exhibit a selectivity for certain tissues. In fact, he developed the concept of distribution of drugs more than anyone had done previously. This specificity had to be explained, and here his work with dyes once again influenced his thinking. He noted that some dyes are thought to become fixed to fabrics by combining with certain constituents of the material to form insoluble salt-like compounds termed 'lakes'. It is possible, he suggested, that animal cells also contain substances which can form lake combinations, and that these substances might differ in different types of cells. Note that although this explanation involves a process of chemical combination, it is quite different from his side chain or receptor theory. The drug is not thought to combine with the cellular protoplasm, but is believed to combine with a substance in the cell, such as an acid, to form an insoluble complex which precipitates out of solution and is thus fixed in the cell. Other dyes were believed to be fixed to fabrics through the formation of 'solid solutions', and Ehrlich suggested that certain drugs might be fixed in cells through a similar process. Once again, this explanation does not

involve the combination of a drug with a receptor.

Ehrlich's doubts about the probability of a 'true chemical combination' taking place between the drug and the cell were shared by many of his contemporaries. At that time, very little was known about the nature of chemical bonding. Chemical combination was generally thought of essentially in terms of covalent and ionic bonds, which are not easily broken, and such concepts as hydrogen bonds and Van der Waals bonds had not yet been developed. It was difficult to reconcile phenomena such as the ease with which many drugs could be washed out of tissues by solvents with the contemporary concepts of chemical bonding. Many pharmacologists of the period therefore preferred to attribute most drug actions to general physicochemical properties of the molecule rather than to the presence of specific structural features which would enable the molecule to combine with the cell.

As late as September of 1906, Ehrlich discussed the question of the affinity of drugs for various tissues or various types of organisms without utilizing the receptor concept. By June of 1907, however, he was able to state:

"I have now formed the opinion that some of the chemically defined substances are attached to the cell by atom groupings that are analogues to toxin-receptors; these atom groupings I will distinguish from the toxin-receptors by the name of 'chemo-receptors'."

What had happened to change Ehrlich's mind about the mechanism of action of drugs? Two important factors appear to have played a significant part in changing Ehrlich's attitude, the work of J. N. Langley and Ehrlich's own studies on drug resistance. In 1913, Ehrlich, looking back at this period, said:

"For many reasons I had hesitated to

apply these ideas about receptors to chemical substances in general, and in this connexion it was, in particular, the brilliant investigations by Langley, on the effects of alkaloids, which caused my doubts to disappear and made the existence of chemoreceptors seem probable to me."

Langley's concept of receptive substances

We must thus turn back to Langley and consider these investigations to which Ehrlich refers. Since 1890, Langley had been investigating the physiology of the nervous system. Langley's experimental work, along with the results of certain other investigators, led him to believe that various substances, like adrenaline, acted directly on muscle cells rather than on the nerve endings in the muscle, as was commonly accepted. It was in studying the effects of nicotine that he obtained his strongest support for this belief, which he discussed in a paper published in 1905. In addition to its well-known paralysing effect, Langley noted, nicotine also has another effect when administered to certain birds. It causes certain muscles to pass into a state of tonic rigidity and to remain in this state for many minutes. This contraction apparently was not due to stimuli transmitted by the nerves, for when all the nerves supplying a muscle were severed the contraction still occurred. This suggested to Langley that nicotine stimulated directly the muscle. If this was true, then the nicotine contraction should not be antagonized by curare, which was believed to exert its paralysing effect on the nerve endings. He found to the contrary, however, that curare had a marked antagonizing effect on the contraction caused by nicotine. Even when the nerve supply to the muscle was severed, curare still antagonized the contracting action of nicotine, and

Langley was forced to conclude that both of these substances were capable of a direct action on the muscle cell.

To explain the observed antagonism between nicotine and curare, Langley returned to his suggestion of 1878, arguing that the two drugs must combine with and compete for the same protoplasmic substance or substances. Langley then added that since neither curare nor nicotine prevents direct stimulation of the muscle (such as electrical stimulation) from causing contraction, they cannot cause paralysis by combining with and blocking the contractile substance. Instead, they must combine with some other constituent of the muscle cell, and Langley called this unknown constituent "receptive substance". He suggested that the normal function of this receptive substance is to transmit a stimulus from the nerve to the muscle. If curare or nicotine combine with the receptive substance, they block the passage of nervous impulses. The contraction caused by nicotine in certain birds was explained by Langley by assuming that the nicotine-receptive substance compound, although blocking nervous transmission, had the ability to stimulate directly the contractile substance (which was unaffected by the drugs).

Langley felt that there was evidence to suggest that many drugs and poisons acted in a similar manner, i.e. by combining with specific receptive substances in the cell. Different cells, he further noted, possess different types of receptive substances. This theory of receptors, first expressed in his initial paper on the subject in 1905 and in his Croonian lecture of 1906, was further elaborated by Langley over the next few years. Further experimental research led him to postulate that the receptive substance of muscle may not actually be a separate compound, but that it could be instead a radical or side-chain of the contractile substance.

He remarked that this concept was along the lines of Ehrlich's side chain theory of immunity, and we must now return to Ehrlich and his conversion.

Ehrlich's chemoreceptor theory

It has already been noted that Ehrlich himself stated that Langley's work was instrumental in changing his mind about drug receptors. It is questionable, however, whether Ehrlich would have been receptive to Langley's views if the studies on drug resistance which were being carried out in his own laboratory at about this same time had not prepared his mind to accept these conclusions. About 1903, Ehrlich and his co-workers began studying the effects of various chemicals on trypanosomes, the microorganisms that cause sleeping sickness. Over the next couple of years, their work and the researches of other investigators uncovered three different classes of compounds which could attack trypanosomes: (1) arsenical compounds (such as atoxyl); (2) azo-dyes (such as trypan red); and (3) basic triphenylmethane dyes (such as fuchsin).

In 1905, the phenomenon of drug resistance was discovered in Ehrlich's lab. His group soon found that the trypanosomes became resistant not to a single chemical but to a whole class of chemical substances. For example, the trypanosome strain which had acquired a resistance to fuchsin was also resistant to other basic triphenylmethane dyes, but it was not resistant to compounds of the other two classes, i.e. arsenicals and azo-dyes. It was apparently in groping for an explanation of this chemical specificity of resistance that Ehrlich was led to accept the concept of drug receptors which Langley had previously put forth. He argued that resistance can be readily explained if one assumes the presence in the trypanosomes of chemoreceptors which combine specifically with a certain type of chemical molecule. The drug must combine with the chemoreceptor in order to act. Ehrlich assumed that the receptors for the arsenicals were different from those for the azo dyes or the triphenylmethane dyes. A trypanosome strain would become resistant to arsenicals if its arsenoreceptors somehow developed a reduced affinity for the arsenic radical. Such strains, however, would still be susceptible to drugs of the other two classes. Once Ehrlich had adopted the idea of chemoreceptors, he rapidly developed the concept further and made it the basis of his theory of chemotherapy.

Ehrlich believed that the chemoreceptors for drugs, like the receptors for toxins, ordinarily are involved in normal physiological functions, hence chemoreceptors exist in all cells. Thus drugs used in chemotherapy are likely to be toxic to the host cell as well as to the parasite cell. One must strive for a substance that has a high affinity for the chemoreceptors of the parasite and a low affinity for those of the host cells. With salvarsan, Ehrlich achieved his most significant therapeutic success.

Ehrlich elaborated on his original chemoreceptor concept by suggesting that a drug may be bound by more than one chemical grouping (i.e. by more than one receptor). He adopted the haptophore-toxophore terminology which he had developed in his immunological work, arguing that a drug was fixed in the cell by one or more haptophore groups, and once it was anchored in this way the toxophore group could then exert its effect. In the case of salvarsan, he postulated, the primary haptophore group is the o-aminophenol group and the toxophore group is the arsenic radical. In Ehrlich's words:

"If, therefore, we poison a spirochaete with salvarsan, at least two different chemical anchorages occur; first, the anchorage of the *o*-aminophenol group which primarily anchors the salvarsan to the parasite. It is only in consequence of this anchorage that, second, the trivalent arsenic radicle is given the opportunity of entering into chemical combination with the arsenoceptor of the cell, and so of exerting its toxic action.''

The impact of the receptor theory of Ehrlich and Langley on pharmacology was very limited in their lifetimes. The science of pharmacology had not advanced far enough by the early decades of the twentieth century to permit a theory which dealt essentially with molecular mechanisms really to be confirmed or denied by experimental means, nor could such a theory yet find significant application in pharmacological research. The interaction between drugs and receptors was not treated quantitatively by pharmacologists until the work of A. J. Clark and J. H. Gaddum in the 1920s and 1930s. And it was not until the 1950s and after that receptor theory became a major area of research interest in pharmacology.

Reading list

1. Ariëns, E. J. (1979) *Trends Pharmacol. Sci.* 1 (inaug. iss.) 11–15.
2. Cuthbert, A. W. (1979) *Trends Pharmacol. Sci.* 1 (inaug. iss.) 1–3.
3. Himmelweit. F. (ed.) (1956–1960) *The Collected Papers of Paul Ehrlich,* 3 Vols., Pergamon Press, Oxford.
4. Langley, J. N. (1878) *J. Physiol.* 1, 339–369.
5. Langley, J. N. (1905) *J. Physiol.* 33, 374–413.
6. Langley, J. N. (1906) *Proc. Roy. Soc. B* 78, 170–194.
7. Parascandola, J. (1974) *Pharm.-Hist.* 16, 54–63.
8. Parascandola, J. and Jasensky, R. (1974) *Bull. Hist. Med.* 48, 199–220.
9. Parascandola, J. (1977) *J. Hist. Med.* 32, 151–171.

This paper is based upon papers published elsewhere (references 7–9 above) where full documentation is provided. The research was supported by grants from the National Science Foundation and the University of Wisconsin Graduate School.

Binding energy and the excitation of hormone receptors

T. J. Franklin

Department of Biochemistry, ICI Ltd., Mereside, Alderley Park, Macclesfield, Cheshire SK10 4TG, U.K.

The 1970s saw an enormously increased interest in the biochemistry of hormone receptors. The availability of many radiolabelled drugs and hormones made receptor binding studies one of the more popular and fruitful activities in biochemical pharmacology. Receptors whose excitation is closely coupled to an easily measured biochemical response, such as increased cyclic AMP synthesis, are especially attractive objects of study because the action of hormones and synthetic analogues on the biochemical responses can be usefully compared with the results of the receptor binding studies. A fascinating and complex picture has emerged of the biochemical mechanisms which couple the excitation of certain receptors to the stimulation of adenylate cyclase[1]. Whilst progress in elucidating the biochemical events associated with the excitation of other types of receptors has been less exhilarating, there is no doubt that receptor excitation is a remarkably effective trigger that initiates the biochemical changes which result in the characteristic biological response.

In complete contrast we know little or nothing about the molecular mechanisms of receptor excitation itself or of the details of the interactions between hormones and their receptors. How do agonists excite their receptors? What is the molecular basis for the efficacy of partial agonists? Why are some competitive antagonists of hormones entirely without agonist activity? Pharmacologists have puzzled over these questions for decades and precise answers await the elucidation of the molecular anatomy of receptors, their specialized microenvironments and the conformational structures of excitatory and inhibitory ligands as they bind to their receptors.

In the meantime we can speculate in the hope that our speculations suggest experiments that may help towards understanding. A major hindrance to scientific progress in the late twentieth century is the compartmentation of ideas that results from extreme specialization. For example, investigators of receptor pharmacology do not, in general, spend much time with enzymologists concerned with the mechanisms of enzyme catalysis. In an attempt at bridge building, I hope to show in this commentary that one of the most important current concepts in molecular enzymology may have major implications for our eventual understanding of the nature of the efficacy of agonists.

Binding energy and enzyme catalysis

A few years ago in a classic review, W. P. Jencks[2] analysed the contribution of the intrinsic free binding energy, ΔG_{int}, released when a substrate binds to the active site of an enzyme, to the efficiency of the catalytic process. He noted that the Michaelis constants, K_m, of many very specific substrates are remarkably high, i.e. between 0.1 and 0.01 mM. If the K_m value measures the affinity of a substrate for its enzyme, such values suggest relatively weak binding, a conclusion apparently at odds with the exquisite specificity of many enzyme–substrate interactions. Jencks resolved this contradiction by proposing that part of the intrinsic free energy of binding is used to 'pay' for the efficiency of catalysis. Transition state theory holds that enzymic catalysis depends essentially upon the ability of enzymes to lower the free energy of activation to achieve the transition state intermediate (Fig. 1). In the absence of the specific enzyme, a reaction such as:

$$A \rightleftharpoons (B) \rightleftharpoons C$$

where (B) is the transition state intermediate between the substrate, A, and the product, C, will be very slow if the energy barrier posed by (B) is high, even though C is thermodynamically favoured. The introduction of the specific enzyme therefore effectively removes the kinetic barrier to the conversion of A to C.

The intrinsic binding energy of the enzyme–substrate complex is used in several ways to reduce the requirement for the free energy of activation to reach the transition state intermediate:

(1) it lowers the requirement for the reduction in translational and rotational entropy involved in the formation of the complex

(2) it may induce conformational changes in the enzyme that are essential for catalysis

(3) it may destabilize the substrate by geometric distortion, desolvation and electrostatic mechanisms.

Elementary thermodynamics tells us that:

$$\Delta G_{obs} = \Delta G_{int} + \Delta G_D - T\Delta S_{int}$$

where
ΔG_{obs} = observed free energy of binding of the substrate (i.e. $-RT \ln K_m$),
ΔG_{int} = intrinsic free energy of binding,
ΔG_D = unfavourable free energy changes due to destabilization of the binding ligand and conformational changes in the enzyme,
$T\Delta S_{int}$ = intrinsic entropy changes, exclusive of solvation effects, associated with the formation of the enzyme–substrate complex.

Since ΔG_D and ΔTS_{int} may be large in relation to ΔG_{int}, the observed affinity of a substrate for its enzyme, measured by the K_m value is apparently substantially worse than it would be if the intrinsic binding were not used to facilitate catalysis.

The different rates at which related substrates are processed by enzymes indicate that only a specific fraction of the intrinsic binding energy is used to facilitate catalysis. For example, related substrates may have similar K_m values but markedly different V_{max} values (i.e. maximum velocity of conversion to product). In contrast, 'poor' substrates with very low values of V_{max} may also have lower K_m values than 'good' substrates. A compound of the latter type would of course be a competitive inhibitor. It is believed that only 'productive' binding interactions induce the correct conformational changes in the enzyme and appropriately destabilize the substrate. In molecular terms, productive interactions involve critical regions of the enzyme and complementary groups on the substrate molecule which are necessary to induce the appropriate changes in both enzyme and substrate to permit catalysis. 'Nonproductive' interactions, in contrast, may cause tighter binding of the substrate (i.e. a lower K_m value) to its enzyme but do nothing to facilitate catalysis.

Application of the binding energy concept to the excitation of hormone receptors

There are obvious parallels between 'good' substrates, 'poor' substrates and enzyme inhibitors on the one hand and agonists, partial agonists and competitive

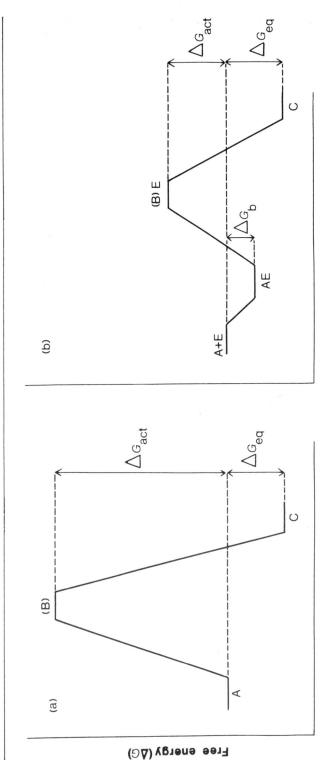

Fig. 1. Free energy diagrams for uncatalysed and enzyme-catalysed reactions. ΔG_{act}: free energy of activation to achieve transition state intermediate (B), ΔG_b: observed free energy of binding for substrate enzyme interaction, ΔG_{eq}: free energy difference between substrate A and product B; note that the enzyme (E) does not change this parameter, i.e. the enzyme does not affect the final equilibrium between A and C. By lowering the free energy of activation the enzyme accelerates the rate of formation of C.

antagonists on the other, although, of course, there is no chemical conversion of hormonal ligands by their receptors. It is generally believed that agonist-induced receptor excitation is associated with a specific conformational change in the receptor protein which is essential for the initiation of the biological response. This conformational change in the receptor corresponds to the substrate-induced conformational change or 'induced fit' that occurs in many enzymes. The ground state P of the receptor or enzyme, which prevails in the absence of the inducing ligand, is assumed to be thermodynamically more stable than the conformationally different or excited state P*. In the presumed equilibrium $P \rightleftharpoons P^*$ the equilibrium constant given by $[P^*]/[P]$ is therefore very small. The excitatory ligand (L) could bring about the conformational change to P* in one of two ways[3]:

(1) By binding specifically, or at least preferentially to P*. The equilibrium is displaced to the right by a simple mass action mechanism i.e.:

$$P \underset{k_{-1}}{\overset{k_1}{\rightleftharpoons}} P^* + L \underset{k_{-2}}{\overset{k_2}{\rightleftharpoons}} P^*L$$

This assumes, of course, that P*L is at a lower energy level than P (Fig. 2).

(2) The ligand binds to P and then *induces* the change to P*, or rather P*L; P*L is assumed to be at a lower energy level than P or PL (Fig. 2).

$$P + L \underset{k_{-3}}{\overset{k_3}{\rightleftharpoons}} PL \underset{k_{-4}}{\overset{k_4}{\rightleftharpoons}} P^*L$$

In both mechanisms the conformational change in the protein is 'paid for' out of the intrinsic free binding energy change when L binds to P* or P. However, the two mechanisms have different consequences for the kinetics of conformational change which may have important physiological and pharmacological implications. The 'induced fit' model of Koshland[3] proposed that the conformational change in an enzyme should follow virtually instantaneously upon the binding of the substrate so that 'induced fit' is not rate limiting in catalysis. On mechanism (1) this requires high values for the forward and reverse isomerization rate constants, k_1 and k_{-1} in order for the ligand to produce a rapid displacement of the equilibrium. Small intramolecular motions in proteins are very rapid and rate constants in excess of 10^3 s^{-1} can easily be envisaged for minor conformational changes[4]. However, conformational changes in proteins may also take much longer than this. For example[4], the dissociation of the subunits of glutamate dehydrogenase induced by GTP or NADH has a forward rate constant of about 60 s^{-1} and the substrate activation of α-chymotrypsin is even slower[4] with a rate constant of $1-3 \text{ s}^{-1}$. There is much evidence to suggest that the energy barrier separating the major conformational forms of many proteins is high enough to impose a considerable kinetic restraint on the exploitation of mechanism (1) to achieve a rapid shift from P to P*[4,5].

Faced with this difficulty it seems reasonable to suggest that by exploiting mechanism (2), some of the intrinsic free energy of binding of an excitatory ligand is used to enhance the forward rate constant, k_4, of the conformational change from PL to P*L. This is achieved by lowering the free energy requirement to reach the hypothetical transition state (P)L between PL and PL* (Fig. 2)[4]. The ligand now plays an *active* role by enhancing the rate at which the active form of the enzyme or receptor is formed. The ability of agonists to accelerate the formation of excited receptors may be important in a receptor system whose excitation involves major conformational changes and consequent protein–protein interactions, e.g. the many adenylate cyclase-linked receptors[1].

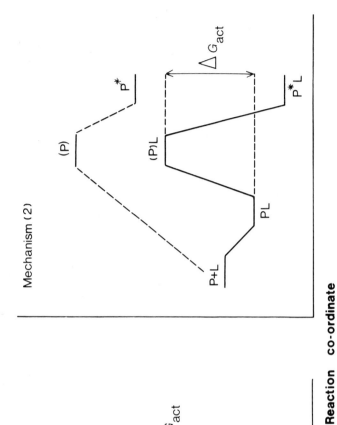

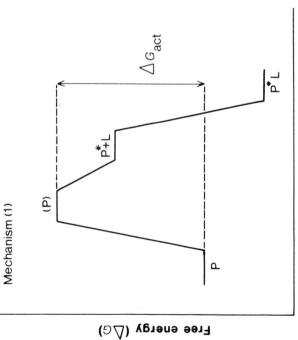

Mechanism (1)

Mechanism (2)

Free energy (ΔG)

Reaction co-ordinate

Fig. 2. Free energy diagrams for alternative mechanisms of agonist-mediated excitation of receptors. In mechanism (1) the agonist L displaces the equilibrium between the ground and excited states of the receptor, P and P*, by binding specifically to P*. In mechanism (2) the agonist actively induces the conformational change in the receptor by binding to the ground state P. In both cases intrinsic binding energy is used to 'pay' for the free energy difference between P and P* but in mechanism (2) it also contributes to a reduction in the activation energy requirement to achieve the transition state intermediate (P) between P and P*.

'Productive' binding energy and the efficacy of agonists

An essential feature of Jencks' model of the contribution of the intrinsic free binding energy to enzyme catalysis is that only 'productive' binding energy can reduce the activation energy requirement to achieve the transition state. The concept of productive and non-productive binding energy can be applied to the problem of the basic nature of the efficacy of agonists and partial agonism (Fig. 3). A popular model[6] to account for the range of efficacies frequently observed within a chemical series of agonists proposes that efficacy is determined by the relative affinities of each compound for the receptor ground state, P (referred to in this model as the antagonist form of the receptor) and the excited (agonist) state, P*.

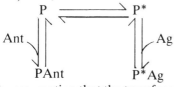

On the assumption that the transformation PAnt → P*Ant is impossible, the model suggests that compounds that bind exclusively to P are pure antagonists with an efficacy of zero while the efficacy of a partial agonist reflects the ratio of its affinities for P and P* as these determine the final equilibrium between the ground and excited states of the receptor[6]. Since P is thermodynamically more stable than P* and P*L is more stable than P, it follows that the intrinsic binding energy released in binding in the agonist, or productive mode, must be used to overcome the unfavourable energy difference between P and P*. The observation that pure antagonists or poor agonists often bind more tightly than corresponding good or full agonists compares directly with the tighter binding of many poor substrates compared with that of good substrates. The essential point here is that the largely non-productive binding of poor agonists is expressed as tightness of

binding whereas the productive binding energy of good agonists is used to bring about the specific conformational change in the receptor protein.

If we consider partial agonism in the light of mechanism (2) for receptor excitation, partial agonists could in principle also be ranked by the *rate* at which they induce the conformational change to excited state of the receptor. We have seen that productive binding energy in mechanism (2) may be used to increase the rate constant k_4 by lowering the activation energy requirement to reach the transition state intermediate (P*L). Because the productive binding energy of partial agonists is a smaller fraction of the total intrinsic binding energy than that of full agonists, the binding energy hypothesis predicts that, for those receptors whose excitation involves mechanism (2), the level of partial agonism

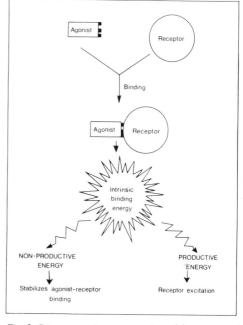

Fig. 3. Diagrammatic representation of the concept of intrinsic binding energy and its partitioning into the energy from 'productive' interactions between the agonist and its receptor, which is used to transform the receptor to its excited state, and 'non-productive' energy which stabilizes the binding between the agonist and receptor.

is not only an expression of the final equilibrium concentration of P*L but should also determine the rate at which P*L is formed. At present this prediction is difficult to test directly because we cannot measure the concentration of P*L in any receptor system. However, Arad and Levitzki[7] have recently shown that the efficacies of several β-adrenergic partial agonists correlate with the rates at which adenylate cyclase is activated and this may reflect the rates at which excited β-receptors are generated by the various agonists.

The molecular basis of productive and non-productive binding energies

Can we relate the concept of intrinsic, productive and non-productive binding energies to the chemical constitutions of agonists, to their receptor interactions and to their efficacies as agonists? Jencks[2] pointed out that intrinsic binding energy is not directly measurable. However, some estimate of the contribution of a substituent group to intrinsic binding energy may be obtained by determining the free energies of binding of a parent molecule A and its substituted derivative A-B to a common receptor. Provided that B does not exert significant steric or inductive effects on the binding of A-B, the difference in the free energies of binding of A and A-B may roughly equal the intrinsic binding energy of B, because the unfavourable entropy change attendant upon binding is essentially the same for A and A-B. Suppose, for example, that in the case of two partial agonists A and A-B, which stimulate adenylate cyclase via a linked receptor, the substituent B also influences efficacy. By measuring the concentrations of the two compounds that half maximally stimulate adenylate cyclase, the rates of activation of adenylate cyclase, and the dissociation constants by the displacement of a radiolabelled pure antagonist, it should be possible to estimate the contribution of

B, or indeed of any other suitable substituent, to both the productive and non-productive binding energies of A-B. The binding studies would have to be performed under appropriate conditions to eliminate the recently reported heterogeneity of agonist binding states[8]. To my knowledge this sort of analysis has not yet been attempted although it could be done with β-adrenergic agents since large numbers of related compounds are available.

Recently there have been reports of measurements of the thermodynamic changes associated with the binding of glucocorticoid agonists to the cytoplasmic receptors of hepatoma cells[9] and of β-adrenergic agents to the β-receptors of turkey red cells[10]. Both studies relied on equilibrium binding techniques and resolved the free energy of binding into enthalpic (ΔH) and entropic ($T\Delta S$) components. Interpretations of the changes in entropy (positive for glucocorticoid agonists, negative for β-adrenergic agonists and positive for β-adrenergic antagonists) in relation to the binding interactions are, however, clouded by the strong possibility that any conformational changes in the receptors consequent upon binding could have contributed to the observed changes in entropy. However, both papers represent interesting attempts to study the energetics of ligand–receptor interactions and they are certain to be the forerunners of many such investigations.

Conclusion

I have outlined the potential importance for receptor research of the concept of the contribution of the intrinsic free binding energy of ligand–receptor interactions to receptor excitation. While the concept is useful in understanding the action of ligands on the equilibria between the ground and excited states of receptors it also suggests that where excitation involves major conformational changes in the receptor with a large requirement for the free energy of activation to reach the transition state intermediate, the efficacy of partial agonists not only reflects the final equilibrium between the ground and excited receptor states but it is also expressed in the rate at which the excited state is formed.

Jencks[2] described the concept of the utilization of binding energy in enzyme catalysis as the 'Circe' effect, alluding to the mythical enchantress of Greek legend who irresistibly attracted men and then turned them into beasts. We can also apply this vivid analogy to hormones and their receptors, although in reverse, for while hormones are drawn into the embrace of their receptors, it is the receptors which are transformed by the encounter.

Acknowledgements

I am grateful to Drs W. P. Jencks and M. I. Page for helpful comments during the preparation of this article.

Reading list

1 Rodbell, M. (1980) *Nature (London)* 284, 17–22

2 Jencks, W. P. (1975) *Adv. Enzymol.* 43, 219–410

3 Koshland, D. E. and Neet, K. E. (1968) *Annu. Rev. Biochem.* 37, 359–410

4 Morawetz, H. (1972) *Adv. Protein Chem.* 26, 243–277

5 Citri, N. (1973) *Adv. Enzymol.* 37, 397–648

6 Ariëns, E. J. (1979) *Trends Pharmacol. Sci.* 1, 11–15

7 Arad, H. and Levitzki, A. (1979) *Mol. Pharmacol.* 16, 749–756

8 Kent, R. S., De Lean, A. and Lefkowitz, R. J. (1980) *Mol. Pharmacol.* 17, 14–23

9 Wolff, M. E., Baxter, J. D., Kollman, P. A., Lee, D. L., Kunz, I. D., Bloom, E., Matulich, D. T. and Morris, J. (1978) *Biochemistry* 17, 3201–3208

10 Weiland, G. A., Minneman, K. D. and Molinoff, P. B. (1979) *Nature (London)* 281, 114–117

How fast do drugs work?

D. Colquhoun

Department of Pharmacology, University College, Gower Street, London WC1 E6BT, U.K.

*Being a very-simplest introduction to those
beautiful methods of reckoning which are
generally called by the terrifying names of
fluctuation and relaxation analysis.*
(With acknowledgement to Silvanus P.
Thompson, 1910)

It is my purpose to discuss some recently
developed methods for the investigation of
the rate of drug action; in particular, to discuss how the methods work, and what they
tell us, rather than to review the many
results that have been obtained by their
use.

The nature of the problem

The question of how rapidly drugs
interact with receptors was probably the
first quantitative pharmacological question
ever to be asked. While A. V. Hill was still a
scholar of Trinity College Cambridge, he
published, in 1909, a remarkable paper[8]. In
this paper he obtained, 9 years before
Langmuir, Langmuir's well-known result
for the binding of a drug (nicotine and
'curari' in his experiments) to adsorption
sites (the 'receptive substance' of muscle).
He derived the relationship between the
amount of binding at equilibrium and the
free drug concentration; he also found the
rate at which this equilibrium binding
should be approached when drug was
added to, or removed from, the tissue bath.
When first approaching the kinetics bus-
iness, there are a number of obstacles to be
overcome. One obstacle is the frightful jargon that has grown up around some aspects

of the subject; another lies in the fact that
this jargon is not always used accurately. A
particularly common, and misleading, mistake is the confusion of the terms *kinetic,
steady-state* and *equilibrium*. The proper
definition of these ideas (in terms of
entropy production) is not very helpful to
most people, but their meanings can be
illustrated by an analogy. Take, for example, the simple Michaelis-Menten mechanism for combination of enzyme (E) with
substrate (S) to produce product (P), i.e.
$E + S \rightleftharpoons ES \rightarrow P$. At equilibrium all the
substrate will be used up; it will be com-
pletely converted to product (because the
last reaction is irreversible). More generally, at *equilibrium* the forward and backward rates for each reaction step will be
equal, and the amount of each reactant will
be constant (not changing with time). A
steady-state is defined by the second condi-
tion alone; for example it is well known
that enzyme reactions will often, after a
short transient period, reach an approximately steady state in which the concentration of enzyme–substrate complex varies
little with time, even though the rate of
association of substrate with enzyme, and
dissociation of substrate from enzyme may
be proceeding at quite different rates. The
study of the *rates* of reactions, e.g. the rate
at which equilibrium is approached is called *kinetics**. The reason that one often
sees equilibrium studies referred to as
'kinetics' lies presumably in the fact that
the equation which describes a simple
enzyme reaction in the (quasi-) steady state

bears a formal resemblance to the Hill--Langmuir equation for adsorption at equilibrium.

Another common bit of jargon is the term *relaxation*. When, for example, the drug concentration is suddenly changed, it is obvious that the occupancy of receptors will change, more or less gradually, until the equilibrium occupancy appropriate to the new drug concentration is attained. Instead of referring to this period of changing occupancy simply as re-equilibration it is often referred to as '*relaxation* of the system (towards equilibrium)'. And the sudden change of concentration is referred to as a concentration *jump*.

Simple binding reactions

Hill considered the simple binding reaction

$$A + R \underset{k_{-1}}{\overset{k_{+1}}{\rightleftharpoons}} AR \qquad (1)$$

in which A represents the drug, and R the receptor. The tendency of an unbound molecule to become bound is measured by the association rate constant, k_{+1}, which is defined by the law of mass action. When the drug concentration is x_A the effective association rate constant will be $k_{+1}x_A$. The tendency of the drug-receptor complex to dissociate is likewise measured by the dissociation rate constant, k_{-1}. The ratio of these rate constants ($k_{-1}/k_{+1} = K$) is the equilibrium constant, which is the concentration of drug at which half the binding sites are occupied at equilibrium[†]. One aim of a kinetic study would be to determine separately the two rate constants, k_{+1} and

---

[*] The term *pharmacokinetics* has, curiously, been sequestered by common usage to describe one small aspect of the subject – the disposition of drugs in the body. This hardly ever reflects the fundamental rate of drug action on receptors and therefore tells us nothing about *how* drugs work, though, as an empirical subject, it is useful in practice.

[†] Hill showed that the occupancy at equilibrium should be $x_A/(x_A + K)$, a result usually, but unjustly, known as the Langmuir equation.

k_{-1}. Hill showed that when drug is added, the fraction of occupied sites (the *occupancy*) should approach its eventual equilibrium value along a simple exponential time course, and that the time constant (τ say) for this exponential curve (i.e. the half-time/0.693) should depend simply on the sum of the rate constants for association and dissociation, thus

$$\tau = 1/(k_{+1}x_A + k_{-1}) \qquad (2)$$

More generally, imagine that the drug concentration is suddenly changed from any original value, to some new value, x_A (this is what might, according to current fashion, be called a concentration jump experiment). The new concentration will correspond to a new equilibrium occupancy, and the transition to this new occupancy should follow an exponential time course with the time constant given above. Notice that the time constant gets shorter (i.e. re-equilibration gets faster) when the drug concentration is high, and that it becomes simply $1/k_{-1}$ when the drug concentration is zero, e.g. the drug is removed and the dissociation of bound drug is followed. Thus, observation of the re-equilibration rate, with a series of different drug concentrations, should allow the determination of the two rate constants, k_{-1} and k_{+1}, in the reaction mechanism.

Behaviour of individual receptors

It is easy to show in a formal mathematical way why re-equilibration should follow an exponential time course, but it is also possible, and much more illuminating, to explain this phenomenon in terms of the way that individual drug-receptor complexes behave. Each drug-receptor complex will not exist for the same fixed length of time; the lifetime of the complex will be *random*. The bonds holding the complex together will be stretching and bending very rapidly (on a picosecond timescale) as the complex is bombarded randomly by water and solute molecules. Occasionally the bonds will be stretched so much that they break, and the drug dissociates. The lifetime of the complex will therefore be

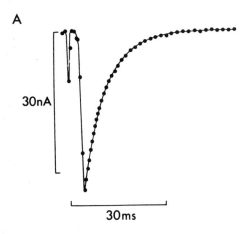

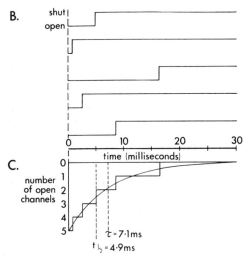

very variable. But this variability of the lifetime will *not* be described by the familiar symmetrical Gaussian (or 'normal') distribution; it *will* be described by a probability distribution with a strong positive skew, the shape of the distribution being, in fact, that of a decaying exponential curve (rather than a bell-shaped Gaussian curve). This is the sort of distribution expected for *random* time intervals (see Ref. 2 for further details). If we consider, for example, the case in which the drug concentration is suddenly reduced to zero, we expect the receptor occupancy to fall (towards zero) along an exponential time course with time constant $1/k_{-1}$. Because we are supposing that there is no drug in solution, it follows that whenever a molecule dissociates from a receptor, the receptor can never be re-occupied, so the fact that the occupancy does not instantly fall to zero when the drug is removed simply reflects the fact that drug–receptor complexes endure for a certain, random, length of time before dissociating. Some dissociate soon after removing the drug; some last much longer. When we observe the exponential decline of occupancy we are simply observing the random (exponentially distributed) lengths of time that elapse before each drug–receptor complex dissociates. The time constant, $1/k_{-1}$ (from equation 2) for the exponential decay of occupancy, is simply the *mean lifetime* of the drug–receptor complex (and the half-time for decay is the *median lifetime*). This argument is illustrated explicitly below (Fig. 1) for a different, but exactly analogous sort of experiment. Similarly, the length of time for which a particular receptor stays *unoccupied* is also random (exponentially distributed). On average it is $1/k_{+1}x_A$ so, as expected, it is shorter for high drug concentrations than for low concentrations.

Why concentration jump experiments rarely work

It has been well known for some time that equilibrium constants ($K = k_{-1}/k_{+1}$)

Fig. 1. A. End-plate current evoked by stimulation of nerve (mean of seven). Frog muscle with membrane potential clamped at −130 mV. The continuous line fitted to the decay phase is a simple exponential curve with a time constant of 7.1 ms (reproduced from Ref. 7, with permission). Inward current is plotted downwards according to the usual convention. About 9000 ion channels are open at the peak.

B. Simulated behaviour of five individual channels that were open at the time (t = 0) at which the acetylcholine concentration had fallen to zero (opening is plotted downwards). They stay open for random lengths of time (mean 1/α).

C. Sum of the five records shown in B. The total number of open channels decays exponentially, as is observed (see A), with a time constant 1/α.

for the binding of competitive antagonist drugs can be obtained by the Schild method, or from direct binding experiments (though measurement of equilibrium *agonist* binding has proved much more difficult). It might be supposed that the sort of experiment described above would have allowed the estimation of the rate constants (k_{+1} and k_{-1}) for the association and dissociation, at least for competitive antagonists. But this, unfortunately, is far from being the case. Ever since 1909 people have continued to ask how rapidly drugs interact with receptors, though they almost all (including Hill) got the wrong answers until about 10 years or so ago. The reason for this lack of success lay, in almost every case, in the obstacle posed by *diffusion* through the tissue. The approach originated by Hill assumed that the drug concentration in the solution immediately adjacent to the receptors was *known and constant.* But in fact the rate of diffusion of drug into the tissue has, in most cases, turned out to be a good deal slower than the rate at which the drug interacts with the receptors so the assumptions of the method were contravened and the observed rate of equilibration reflected nothing but the rate of diffusion into the tissue. Of course this problem did occur to many workers; so, in some cases, they checked their results by seeing whether the ratio of the dissociation rate constant and association rate constant, which were inferred from the observed equilibration rate, did actually agree, as they should, with the value of the equilibrium constant (which could be *independently* determined by the Schild method, or from binding experiments). Often quite good agreement was found, and this seemed to suggest that the rates of association and dissociation must be about right. This sounds very reasonable, but unfortunately it is not. The reason lies in the fact that in many cases the rate of diffusion of drug into the tissue is strongly dependent on the extent of the binding of the drug.

When a molecule of drug becomes bound, it is no longer free to diffuse towards the centre of the tissue so if a drug can be bound by the tissue it will diffuse more slowly than a molecule that cannot be bound. Furthermore, the extent of this slowing will depend on, among other things, the fraction of the drug that is bound, and hence on the *equilibrium constant* for binding of the drug. Thus it turns out that the diffusion rate of the drug may be dominated by its equilibrium constant for binding so it is not surprising that the observed rates of onset and offset of the drug can tell us something about its equilibrium constant, even when these rates are quite unrelated to the true rates of association with, and dissociation from, the receptors. Diffusion can mimic the kinetic behaviour predicted by equation 2 with alarming precision, hence the frequent lack of success (see for example, Ref. 6). Other methods are clearly needed.

Methods that do work

During the last 10 years or so, some methods have been developed that really can measure rates. Responses to drugs can be roughly classified into fast mechanisms which equilibrate in a few milliseconds (generally those involved in fast synapses, e.g. the nicotinic acetylcholine receptor or GABA responses), and, on the other hand, slow mechanisms which take more than 0.1 sec to equilibrate (such as responses of smooth muscle or gland cells to muscarinic acetylcholine or β-catecholamine receptors). Curiously enough, much more is known about the rates of the faster responses than of the slower ones: this is because (a) faster responses can be followed by (relatively rapid) electrical measurements (because the effect of the agonist is to open ion channels), whereas the slow ones are often still not slow enough to be followed by the appropriate (much slower) chemical measurements, and (b) the mechanisms of fast responses seem (so far) to be much simpler than those of the slower responses, so observations on fast reactions are easier to interpret.

A simple mechanism for fast responses

A convenient mechanism with which to illustrate many of the principles involved is that first postulated by Castillo and Katz in 1957 for the nicotinic acetylcholine receptor at the neuromuscular junction. An agonist molecule (A) binds to an inactive (i.e. ion channel shut) receptor (R), to form a complex (AR) which may then isomerize to the active state (AR*) in which the ion channel is open.

$$
\begin{array}{cc}
\text{shut} & \text{open} \\
\overbrace{R \underset{k_{-1}}{\overset{k_{+1}}{\rightleftharpoons}} AR}_{\text{vacant}} & \overbrace{\underset{\alpha}{\overset{\beta}{\rightleftharpoons}} AR^*}^{} \\
& \underbrace{}_{\text{occupied}}
\end{array} \quad (3)
$$

Electrical measurements allow us to follow the number of ion channels that are open (AR), though they cannot tell us how the shut channels are divided between the two shut states, R (vacant), and AR (occupied). The chemical (mass action) rate constant for each step is indicated on the appropriate arrow in equation (3). The rate at which this mechanism approaches equilibrium can be derived by standard chemical kinetic methods, but, just as above, it is more enlightening to think in terms of the behaviour of single receptors. For example, the length of time for which an ion channel stays open (its *open lifetime*) will vary randomly, as shown in Fig. 3, with an average duration of $1/\alpha$. Similarly the length of time spent in the occupied but shut state, AR, will*, on average, be $1/(\beta + k_{-1})$.

---

* Notice that the mean lifetime of any state depends on the sum of the rate constants for all the pathways by which that state can be left. And the variability of the lifetime will, as in the case of the occupied lifetime discussed above, be described by an exponential probability distribution (see Refs 4 and 5 for details).

A special case

The mechanism in (3) postulates that the system can exist in three distinct states. It is, however, commonly believed that the first (drug binding) reaction is very much faster than the second (channel opening) reaction. If this were the case (which is probably not strictly true: see below) then the mechanism in (3) could be simplified to:

$$
\text{shut channel (R or AR)} \underset{\alpha}{\overset{\beta'}{\rightleftharpoons}} \text{open channel (AR)}
$$

$$(4)$$

The true opening rate, β, has been replaced here by the effective opening rate, β', which (unlike β) increases with agonist concentration so $1/\beta'$ is simply the mean lifetime of the shut state†. Just as in the simple binding reaction discussed above, the mechanism in equation (4) would be expected to approach equilibrium along a simple exponential time course; the time constant (half time/0.693) for this exponential will again depend on the sum of the forward and backward reaction rate constants, thus:

$$\tau = 1/(\alpha + \beta') \quad (5)$$

= mean open time × fraction of channels shut (at equilibrium).

Thus, when the agonist concentration is low (so the opening rate β' is much slower than the shutting rate, α, and the fraction shut is near to one) the time constant is approximately equal to $1/\alpha$, the mean open channel lifetime.

Two types of method

There are two classes of method for measuring rates. These are as follows.

---

† The effective opening rate is the true rate multiplied by the fraction of shut channels that are occupied (because only the occupied shut channels are *capable* of opening), so, from equation (2), $\beta' = \beta x_A/(x_A + K_A)$. The fraction of channels open at equilibrium is $\beta'/(\alpha + \beta')$.

(1) *Relaxation methods.* The obvious method, exemplified above, is to change ('jump') some variable (drug concentration in the example above) so that the system is no longer in equilibrium (it is 'perturbed'), and then to follow the rate at which the system approaches (*relaxes towards*) its new equilibrium.

In the case of enzymes, the position of equilibrium can be changed by sudden changes (jumps) in temperature or pressure, (as well as concentration), but these methods are not (yet) feasible for most drug responses. The method of this class that has been most successful is to change the position of equilibrium by a sudden change in *membrane potential* (a *voltage jump*). This, of course, was the method used by Hodgkin and Huxley (long before the rather pompous expression 'voltage jump relaxation' came into common use), for their classical work on axons; examples of its use with drugs are discussed below.

In all of these methods a *sudden* change of concentration (or temperature, or membrane potential) is desirable so that the concentration (etc.) stays constant during the ensuing re-equilibration.

(2) *Equilibrium methods.* The second class of method can be (and usually is) applied when everything is at equilibrium. It sounds odd, at first, that *rates* can be measured when the system is in *equilibrium*. But it is quite clear that this can be done, at least if we can see individual events[11]; for example the lifetimes of individual ion channels can be measured from records such as that in Fig. 3, and the average of these lifetimes can be interpreted (if a mechanism like that in equation 4 is appropriate) as $1/\alpha$. We can therefore obtain an estimate of the chemical rate constant, α, for channel closure. This method works because the term *equilibrium* refers only to the *average* characteristics of the system. If, for example, half the ion channels are open at equilibrium, this does not mean that every ion channel is half open all the time, but that each ion channel

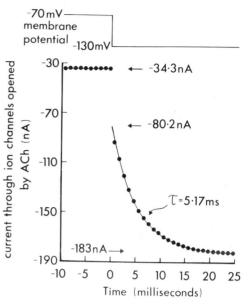

Fig. 2 *Re-equilibration after a sudden change in membrane potential. Acetylcholine, applied to a frog muscle end-plate, produces an inward current (plotted downwards) of 34.3 nA at a membrane potential of −70 mV. After imposition of a sudden change of membrane potential to −130 mV at zero time (as shown at the top), the current changes gradually to its new equilibrium value of 183 nA. The time course of this re-equilibration has been fitted with a simple exponential curve with a time constant of 5.17 ms.* (Colquhoun, Dreyer and Sheridan; unpublished).

is fully open for half of the time, and fully shut for the rest of the time, so an individual ion channel is, in a sense, *never* at equilibrium (indeed the term is not really defined for single channels).

In Fig. 3 only one channel is open at a time; if many are open simultaneously, individual openings and shuttings may not be distinguishable, and the record will merely look noisy as a result of the fact that the number of open channels is not the same from moment to moment. But similar information can still be extracted from this *noise* by the methods discussed below.

These various methods will now be illustrated.

The methods in practice

(1) *Re-equilibration after a sudden change of concentration (concentration jump relaxation).*

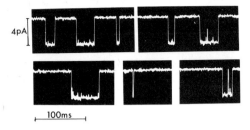

Fig. 3. Examples of the currents that flow when individual ion channels open under the influence of the
cholinomimetic agonist suberylcholine (20 nM) (opening is shown as a downward deflection). Notice the con-
stancy of the amplitude (3.6 pA), but great variability of
the length of time for which the channel stays open (see
text). Two openings occur in quick succession in the last
(and perhaps the fifth) events shown. Records are from
perisynaptic receptors of frog cutaneus pectoris muscle
at a membrane potential of −130 mV (B. Sakmann and
D. Colquhoun; unpublished observation).

The reason why attempts to produce
concentration jumps rarely work were discussed above. One case in which this
method did, almost certainly, work was
described by Schwarz, Ulbricht and
Wagner in 1973. They were able to perfuse
single Ranvier nodes of myelinated axons
so fast that the concentration of drugs at
the site of action could be changed within a
second or so. They found that tetrodotoxin
dissociated from the sodium channels in
the node with a time constant of about 70
seconds, much longer than the time for
removal of tetrodotoxin from solution.
Therefore k_{-1} could be estimated as $1/(70$
sec$) = 0.014 \ sec^{-1}$.

An ingenious method of producing a
rapid concentration change has been used
by Lester[10]. He used a light-activated agonist, the concentration of which could be
·rapidly changed by a very brief flash of
light.

Another way to produce a rapid change
in the concentration of agonist is to use
acetylcholine released from a nerve ending
(either spontaneously or by nerve stimulation). The acetylcholine concentration in
the synaptic cleft rises quickly to a peak,
ion channels open, but then, under some
conditions, the acetylcholine disappears

(because of hydrolysis and diffusion) very
rapidly so that shortly after the peak
response, there is no acetylcholine left[1].
Thus, to a fair approximation, there is a
jump in acetylcholine concentration from a
high value to zero, so useful kinetic information can be obtained from the rate at
which ion channels close after nerve stimulation, or spontaneous release of a quan-
tum of transmitter. Fig. 1B shows schematically the history of five individual ion
channels that were open at zero time (it
does not matter, for our purpose, when
they originally opened; see Ref. 2). They
stay open *on average*, for a time $1/\alpha$, the
mean lifetime of the open ion channel. But
individual channels stay open for a randomly variable (exponentially distributed)
length of time, and, because the agonist
concentration has fallen to zero, once a
channel shuts it can never re-open. Therefore the total number of open channels,
shown in Fig. 1C, is expected to decay
exponentially (with a time constant $1/\alpha$)
until all are shut. This, as illustrated in Fig.
1A, is just what does happen. And therefore, insofar as the theory (based on a
mechanism like that in equation 4) is correct, the decay rate of end-plate currents
gives an estimate of the mean open channel
lifetime*, and hence of the shutting rate
constant, α.

(2) *Re-equilibration after a sudden change
in membrane potential (voltage jump relaxation).*

The opening and shutting of ion channels in axons is controlled by membrane
potential. But at chemical synapses it is
agonists that control the opening and shutting of ion channels. However, even drug-
operated ion channels are usually somewhat sensitive to membrane potential, as
well as to agonists, so a sudden change in
membrane potential can change the posi-

---

* With greater formality (and less illumination) this
follows from equation 5 because the opening rate, β' is
zero if there is no agonist present.

tion of equilibrium; for example a fixed acetylcholine concentration opens more channels when the end-plate is hyperpolarized than when it is depolarized. This is illustrated in Fig. 2, which shows the current that flows through ion channels opened by acetylcholine in frog motor end-plate. At equilibrium the inward current is 34.3 nA (at a membrane potential of −70 mV). Then the membrane potential is changed suddenly to −130 mV. There is an *immediate* increase in current to 80.2 nA (because there is more voltage to drive it through the channels that are already open). Then comes the interesting bit; there is a gradual increase (which follows a simple exponential time curve) in the current as more ion channels open, until eventually the equilibrium current (183 nA) at the new membrane potential (−130 mV) is reached. If we interpret the result in terms of the mechanism in (4), the time constant (5.17 millisecs in Fig. 2) for the experimental re-equilibration must simply be (from equation 5) $1/(\alpha + \beta')$. So, in principle anyway, such measurements (at various agonist concentrations) can give information about α, β and the equilibrium constant for agonist binding (see, for example, Refs 13 and 14). And, as before, at low enough agonist concentration, the time constant will be close to the mean open channel lifetime.

(3) *Observation of single ion channels.*

Channels opening under the influence of suberylcholine are shown in Fig. 3. As described earlier, the mean channel lifetime can be measured directly from such records. However, although each current originates from one ion channel only, it is not necessarily the *same* channel that opens each time an opening is seen on a record like in Fig. 3. Therefore the average interval *between* openings (which is, in principle, even more informative than the open periods) cannot be interpreted easily except under special conditions where it is known that every opening originates from

the *same* channel (see Ref. 15), in which case the mean interval between openings gives an estimate of $1/\beta'$ (see above).

As well as giving information about kinetics, single channel records also show directly the amplitude of current pulses (about 3 pA), and hence the conductance of a single open ion channel, which is commonly about 30 pS (i.e. resistance about 33 000 megohms; one Siemen = one reciprocal ohm). This is information that cannot be obtained by the 'relaxation' methods though it can be estimated from noise analysis.

(4) *Observations of noise, or fluctuations, in the response to agonists.*

When many ion channels are simultaneously open, the individual step-like tran-

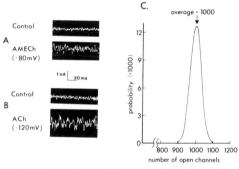

Fig. 4. *Fluctuations of the current through rat diaphragm end-plate, in the absence of any agonist (top) and in the presence (below) of agonist, as follows.*
(A) acetylmonoethylcholine (at a membrane potential of −80 mV). The fluctuations have a standard deviation of 0.20 nA, and were superimposed on a response (mean current) of 27 nA. At equilibrium about 13,000 ± 190 (2 standard deviations) were open.
(B) acetylcholine (at a membrane potential of −120 mV). The fluctuations have a standard deviation of 0.36 nA, and were superimposed on a response (mean current) of 64 nA. At equilibrium about 21,000 ± 230 channels were open. (Reproduced from Ref. 16, with permission).
(C) The predicted variability in the number of open ion channels. Calculated from binomial distribution for the case where there are one million channels, each with a chance of 1 in a 1000 of being open, so the number of channels that are open on average (i.e. at equilibrium) is 1000 (see text), with a standard deviation of 31.6. For about 5% of the time the number that is open deviates from the average by more than two standard deviations (i.e. is less than 937, or more than 1063).

sitions shown in Fig. 3 may (depending on the recording method) not be distinguishable. But they are there, and they give rise to the noisiness of the record in the presence of agonists shown in Fig. 4A and B.

It has already been mentioned above that the term equilibrium refers only to *average* behaviour. Imagine that there are one million ion channels in the experimental preparation, and that we apply enough agonist to open, at equilibrium, one channel in a thousand. Therefore at equilibrium there are, on *average*, 1000 ion channels open. But *all* of the channels are opening and shutting at random and there is not *exactly* one thousand open at any particular moment. Suppose that an instantaneous picture could be taken of all the channels at a particular instant, and the state (open or shut) of each one noted. This would be exactly analogous to tossing a penny a million times and noting, each time whether it came down 'heads' (channel open) or tails (channel shut). (In this case it is a very biased penny, that comes down heads only once in 1000 throws on average.) Clearly every time the experiment was repeated (an instantaneous picture taken, or a million tosses performed), we would not expect to see *exactly* 1000 open channels (or to get exactly 1000 'heads'). The problem can be treated by standard coin tossing theory*, and the expected distribution of the number of open channels is shown in Fig. 4C. It has a standard deviation of 31.6 and a roughly Gaussian form. So we expect that 95% of the time, the number of open channels will be within two standard deviations (i.e. 63.2) of the mean and we should therefore say, not that there are 1000 channels open at equilibrium, but that there are 1000 ± 63.2. These moment–to–moment fluctuations (of about ±6 percent in this example) in the

number of open ion channels are quite big enough to see, and they are what gives rise to the fluctuations, or noise, shown in Fig. 4A, B[9].

How, then, can useful information be extracted from the rather unpromising-looking signals shown in Fig. 4A, B? There are two things that we can measure, the amplitude of the noise, and its frequency. Let us deal with these in turn.

Noise amplitude. A conventional way to measure the amplitude of a fluctuating current (e.g. the alternating mains supply) is to specify the rms (root mean square) current, which is nothing other than the standard deviation of the current, calculated in the ordinary way from a series of values of the current measured at different instants. This amplitude (standard deviation) is, for example, clearly bigger for the record in Fig. 4B than for that in Fig. 4A. The standard deviation of the current is a direct measure of the standard deviation of the number of open channels which was discussed above. If we divide the square of the observed standard deviation by the mean current, we obtain an estimate of the current through a single ion channel (around 3pA) and hence an estimate of the channel conductance (as long as few channels are open and a mechanism like equation 4 is valid†).

Noise frequency. It is obvious by eye that the noise in Fig. 4A contains higher frequencies than that in Fig. 4B. It is therefore natural to ask whether anything interesting can be learned from the frequency, and, if so, how it can be measured. The answer to the first question is yes; we can learn about the rate at which channels open and shut.

---

* The binomial distribution with $N = 10^6$ channels, $p = 0.001$ (probability of being open), gives the mean number of open channels as $Np = 1000$, with standard deviation $\sqrt{Np\,(1\text{-}p)} = 31.6$.

† Under these conditions $Np(1 - p) \simeq Np$ so the variance of the total conductance is approximately $\gamma^2 Np$ (where γ is the conductance of a single channel) whereas the mean conductance is γNp. Their ratio is simply γ.

The answer to the second question follows. Fig. 5A shows, schematically, the opening and shutting of five individual ion channels (mean channel lifetime = 5 ms, say), and Fig. 5B shows the sum of these records. The fluctuating record in Fig. 5B is just the sort of noise illustrated in Fig. 4A,B (though the individual steps are too small to be distinguished in the latter). Now at time zero it is seen that there are four ion channels open (channels number 1, 2, 4 and 5). If we look a bit later on, say 1.0 ms later (as marked in Fig. 5B), there are still four channels open; not surprisingly, they are the same four as were open originally, because, in a short time like 1.0 ms it is quite likely that none of these channels will have shut yet (and that no others will have opened). It is therefore to be expected that the current at any given moment will be very similar to the current at a short time later if, by short, we mean *short relative to the channel lifetime*. In other words the current at any given moment will be highly correlated with the current 1.0 ms later. If we calculate an ordinary correlation coefficient between these two quantities we expect that it will be very good, i.e. near unity, as shown in Fig. 5C. Now consider what happens if, rather than waiting 1.0 ms, we wait 15 ms (see Fig. 5B). This is quite long compared with the mean open lifetime, so it is not surprising to see that all the channels that were originally open have shut in the intervening 15 ms, and some others have opened (or the same ones re-opened). Clearly there is no reason to suppose that, just because the current was above average, at zero time, it will still be above average after 15 ms. It could be anything; in other words if we calculate the correlation between the current at a given moment, and the current a long time (relative to the channel lifetime) later, we expect to find a low correlation, as plotted in Fig. 5C. Naturally, at intermediate times comparable to the channel lifetime (5 ms in Fig. 5B), we expect some of the original

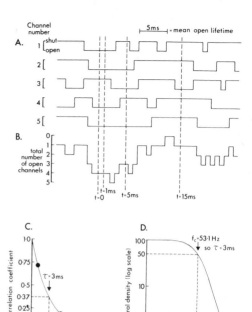

Fig. 5. Explanation of the analysis of the frequency characteristics of fluctuations in the current elicited (i.e. in the number of ion channels opened) by agonists.

A. Simulated behaviour of five individual ion channels (opening is plotted downwards). They are opening and shutting at random with a mean open lifetime of 5 ms.

B. Sum of the five records shown in A. The total number open (and hence the total current) shows fluctuations of the sort that give rise to observations like those in Fig. 4A,B. An arbitrarily chosen line marks zero time, and the times 1, 5 and 15 ms later are also marked with vertical lines.

C. The correlation analysis of noise. Two observations separated by 1 ms are likely to be highly correlated (as shown by the dot). The correlation will be less when the separation is 5 ms, and there is hardly any correlation with a separation of 15 ms. The correlation (according to the mechanism in equation 4) dies out along the exponential curve shown, with a time constant of $\tau = 1/(\alpha + \beta')$. The time constant is 3 ms in the example shown; it is less than the mean open channel lifetime (5 ms), because, as is clear from A, the channels are open for a substantial part of the time. (From equation 5 it follows that the equilibrium fraction of shut channels is 0.6, of open channels is 0.4, $\alpha = 1/(5 \text{ ms}) = 200 \text{ s}^{-1}$ and $\beta' = 133.3 \text{ s}^{-1}$: thus $\tau = 1/333.3 = 3$ ms).

D. Presentation of the same analysis of noise as a power spectrum. The ordinate is normally in units of Amp²/Hz, but is given in arbitrary units here. The amount of noise is halved (from 100 to 50) at 53.1 Hz, so $\tau = 1/(2\pi \times 53.1) = 3$ ms, exactly as found in C.

channels to be still open (e.g. channels 2 and 5 in Fig. 5A), and others to have shut. We therefore expect an intermediate degree of correlation as shown in Fig. 5C. When a graph of correlation (usually called autocorrelation in this context) is plotted against time interval, as in Fig. 5C, the reader may well not be surprised, by this time, to find that theory (for mechanisms like equation 4) predicts that the graph will have the form of an ordinary decreasing exponential curve, and that the time this curve takes to decay is related to the mean open channel lifetime. In fact, Nature, with incredible elegance, has contrived that the time constant for this decay is simply $1/(\alpha + \beta')$, exactly the same value (see equation 5, and above) as found from measurements of re-equilibration.

But, the reader may comment, correlation graphs, like that in Fig. 5C, are only rarely seen in experimental papers. This is, perhaps sadly, true. There is another, and more common way of plotting the analysis of the frequency of noise. It is exactly equivalent to the correlation graph just shown and may be regarded simply as a transformation of it (a bit like transforming an exponential curve by plotting it semilogarithmically); rather than plotting time along the abscissa, reciprocal time (i.e. frequency) is plotted. And the ordinate reasonably enough, gives the amount of noise* at each frequency. Such a graph is shown in Fig. 5D. It is called a power spectrum (or, better, as a spectral density function). The curve drawn in Fig. 5D is called a Lorentzian curve; it is merely the equivalent (on a spectral density graph) of the

exponential curves shown in all the other methods of analysis so far. The graph is flat at low frequencies, but eventually bends downwards. And, clearly this bend would be expected to occur at higher frequencies for noise such as that in Fig. 5A than for that in Fig. 5B, because the former clearly contains wobbles of higher frequency than the latter. The position of the bend is usually measured by the *cut-off frequency* (f_c), that is, the frequency at which the amount of noise is half that at low frequencies. As would be expected from the equivalence of the presentations in Figs. 5C and D, this frequency is very simply related to time constant, τ, found from Fig. 5C (by the relationship $\tau = 1/2 \pi f_c$). Both analyses of noise frequency, if interpreted according to mechanism (4), give us a value for $1/(\alpha + \beta')$.

More complicated cases

So far we have dealt only with the simplest case of the kinetics of a single agonist, the binding of which is rapid (see equation 4).

In fact it is quite likely that binding is not as rapid as required by equation 4; sometimes two or more openings may occur in quick succession during the time the receptor stays occupied (this may be the cause of the multiple openings noted in Fig. 3). If this happens, the interpretation of kinetic experiments gets more complicated.

In the presence of certain antagonists, the simple exponential curves (shown in Figs 1, 2 and 5) become sums of two or more exponentials (and, correspondingly, the power spectrum becomes the sum of two or more Lorentzians). These kinetic changes, which correspond to the occurrence of openings in clusters, have given valuable information about how some antagonists work. Their interpretation proceeds along much the same lines as have just been discussed for simple problems but the details are beyond the scope of this article. An account will be found in reviews (e.g. Refs 3 and 12).

* More precisely, it gives the variance (standard deviation squared) of the noise at each frequency. It could be found (approximately) by feeding the noise into a filter that was set to pass only frequencies between, say, 9.5 and 10.5 Hz. The standard deviation of the signal coming out of the filter would be squared and plotted against 10 Hz on the abscissa. If this were repeated with different filter settings a curve like that in Fig, 5D could be constructed. This description also explains why the area under curve is equal to the variance of the original current.

Reading list

1 Anderson, C. R. and Stevens, C. F. (1973) *J. Physiol. (London)* 235, 655–691

2 Colquhoun, D. (1971) *Lectures on Biostatistics*, Clarendon Press, Oxford

3 Colquhoun, D. (1979) in '*The Receptors: a comprehensive treatise*' (O'Brien, R. D., ed.), Plenum Press, New York

4 Colquhoun, D. and Hawkes, A. G. (1977) *Proc. Roy. Soc.* B199, 231–262

5 Colquhoun, D. and Hawkes, A. G. (1981) *Proc. Roy. Soc.* B211, 205–235

6 Colquhoun, D. and Ritchie, J. M. (1972) *Molec. Pharmacology* 8, 285–292

7 Colquhoun, D. and Sheridan, R. E. (1981) *Proc. Roy. Soc.* B211, 181–203

8 Hill, A. V. (1909) *J. Physiol. (London)* 39, 361–373

9 Katz, B. and Miledi, R. (1970) *Nature (London)* 226, 962–963

10 Lester, H. A. (1977) *Scientific American* 236, 105–118

11 Neher, E. and Sakmann, B. (1976) *Nature (London)* 260, 799–802

12 Neher, E. and Stevens, C. F. (1977) *Ann. Rev. Biophys. Bioeng.* 6, 345–381

13 Osterrieder, W., Noma, A. and Trautwein, W. (1980) *Pflügers Archiv* 386, 101–109

14 Sakmann, B. and Adams, P. R. (1979) in *Advances in Pharmacology and Therapeutics Vol. 1 Receptors* (Jacob, J., ed.), pp. 81–90, Pergamon Press, Oxford

15 Sakmann, B., Patlak, J. and Neher, E. (1980) *Nature (London)* 286, 71–73

16 Colquhoun, D., Large, W. A. and Rang, H. P. (1977) *J. Physiol. (London)* 266, 361–395

Desensitization

D. J. Triggle

Department of Biochemical Pharmacology, School of Pharmacy, State University of New York at Buffalo, Buffalo, NY 14260, U.S.A.

The phenomenon of diminished response during, or subsequent to, the initial action of a drug (variously described as tachyphylaxis, tolerance, refractoriness, subsensitivity or desensitization) has long been observed in a wide variety of cellular systems. 'Desensitization' is increasingly used as a global term to describe the loss of cellular sensitivity, subsequent to agonist treatment, regardless of mechanism, and will be used as such in this article.

In 1898, Tigerstedt and Bergman described diminished pressor responses during repeated renin administration and Dale in 1913 observed desensitization in an entirely different system, anaphylactic reactions in guinea-pig smooth muscle. The ubiquitous occurrence of both specific and non-specific desensitization in systems from bacterial to mammalian, suggests that desensitization represents an important regulatory component of the homeostatic capacity of cellular recognition processes. Recent developments in this area have clearly indicated the generality of the conclusion that conditions which decrease or increase agonist–receptor interaction, either chronically or acutely, result in opposing alterations in effector sensitivity[1]. This may be seen as an extension of Cannon's law of denervation.

Desensitization and receptor events

With increasing awareness of the events initiated by agonist–receptor interactions it has become apparent that a number of mechanisms probably contribute to the desensitization process. Specific desensitization is commonly attributed to events at the receptor itself and non-specific

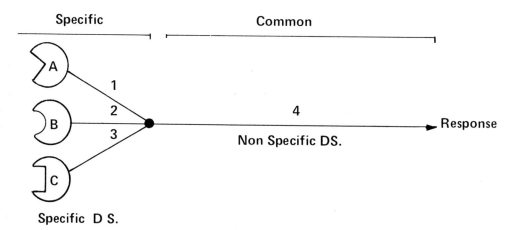

Fig. 1. Specific and non-specific desensitization in a multiply sensitive system. A single response is shown as being initiated through a common sequence (4) initiated by three specific pathways (1, 2, and 3) linked to the specific recognition sites A, B and C.

TABLE I. Consequences of receptor activation

Formation and activity of the A-R complex.
(a) A-R: coupled to ion channel (X) or other amplification unit.
(b) A-R-X: initiates response,
(c) A-R-X: X 'turns off', system desensitizes,
(d) X: products of X activity inhibit function of A-R-X, system desensitizes.
Fate of A-R-X complex.
(a) A-R: becomes uncoupled from X, system desensitizes.
(b) A: becomes uncoupled from R-X, system desensitizes.
Fate of A-R complex.
(a) A dissociates, R becomes reavailable
(b) A dissociates, R is inactive, system desensitizes
(c) A-R is shed from surface – receptor loss, system desensitizes
(d) A-R internalized – receptor loss, system desensitizes

desensitization in multiply sensitive systems to events distal to any specific receptor component (Fig. 1). However, it is clear that non-receptor events also contribute as with increased metabolism in barbiturate tolerance and decreased catecholamine stores in tachyphylaxis to indirectly acting sympathomimetic amines. Some general consequences of receptor activation and their potential role in mediating desensitization are summarized in Table I. There appears to be a major difference between

acute desensitization, which develops rapidly and is usually rapidly reversible, and chronic desensitization which develops slowly and is less rapidly reversible and is often accompanied by true receptor loss (down-regulation) rather than the loss of receptor function which is typical of acute desensitization. The process of receptor loss has been described for a number of receptor systems and involves a dominant role for receptor clustering in specific membrane areas termed coated pits – the membrane analog of galactic black holes – from which endocytosis occurs with degradation or delivery to intracellular organelles of the hormone[2].

Mechanisms of desensitization

Several factors must be considered in any general consideration of the mechanisms of desensitization. A clear distinction must be made between acute and chronic phenomena, the latter being much more likely to involve a true receptor loss. Careful consideration must be given to the experimental conditions since the desensitization process in whole cells is different in a number of instances from the process in membrane fragments. The specificity of the process must be considered both in acute and chronic states since an initially

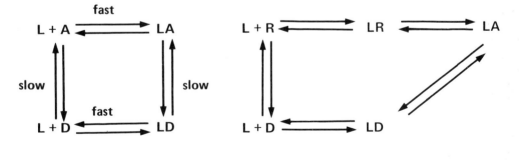

a   b

Fig. 2. Cyclic schemes for desensitization of the nicotinic acetylcholine receptor. (a) The original Katz-Thesleff equilibrium scheme in which the rapid interaction of ligand (L) with receptor in state A gives the active complex LA, subsequently forming the desensitized state LD. (b) A modification in which the initial rapid formation of the inactive ligand receptor complex LR is shown which then undergoes conversion to state LA with transient opening of the associated ion channel.

specific desensitization process may become subsequently non-specific. Finally, consideration must be given to the occurrence of dispositional changes including ligand metabolism and substrate depletion.

In 1957 Katz and Thesleff[3], using iontophoretic application of agonists to skeletal muscle nicotinic receptors, were able to quantify a rapid desensitization and recovery process, the kinetics of which were consistent with a cyclic scheme, involving interconversion of active (A) and inactive high affinity (B) forms of the receptor (Fig. 2a). When written in the equilibrium form as shown, this scheme implies that there exists, even in the resting state, a distribution of receptors between the active and desensitized states. If into this scheme is inserted an additional step to represent the associated ionophore activation step then the receptor activation–desensitization process may be viewed as implying the existence of three receptor states, resting, activated and desensitized, with the associated ionophore being open only in the activated state (Fig. 2b). This concept appears consistent with kinetic measurements indicating the existence of low (R), medium (A) and high (D) affinity states, with the ability of some antagonists ('metaphilic agents') to bind preferentially to the D state and with the ability of Ca^{2+} and local anesthetic to facilitate desensitization by promoting formation of the A or D states[4,5]. In this perspective receptor activation may be viewed as a transient activation of the A state and desensitization (D) as the final equilibration, but the extent of this interconversion will be agonist-dependent. Desensitization subsequent to chronic treatment with cholinergic agonists has been demonstrated in this system and, in common with many other systems, is accompanied by receptor loss in addition to the reduction in receptor sensitivity.

Desensitization of several hormone-sensitive adenylate cyclase systems has been described. Here also, desensitization is an agonist-induced process, is prevented by antagonist treatment and is absent in mutant cells lacking adenylate cyclase, consistent with the thesis that desensitization is mediated through receptor activation[6]. However, there is clearly no single mechanism for receptor desensitization in these systems.

Isoproterenol (10^{-4}M, 1–3 h) mediated desensitization of the frog erythrocyte results both in actual receptor loss ($c.$ 75%) and an impairment of the ability to form high affinity agonist–receptor complexes[7]. This may account for the common observation in desensitized adenylate cyclase systems that the loss in activity is proportionately greater than receptor loss. Significant differences are noted between desensitization in erythrocytes and erythrocyte membranes[8]. In the latter preparation isoproterenol desensitization is very rapid ($t_{1/2}c.$ 10 min), is accompanied by a smaller reduction in AC activity and probably represents only the process impairing the function of the active hormone–receptor–effector complex consistent with the desensitization-protecting effect of guanyl nucleotides.

Other adenylate cyclase systems present contrasting patterns of desensitization. In human astrocytoma cells desensitization appears to be mediated via cAMP[9], presumably through a protein kinase-directed phosphorylation, and in rat astrocytoma cells cAMP-dependent desensitization is dependent on new protein synthesis[10] presumably reflecting the generation of an endogenous inhibitor. Of particular interest are multiply hormone-sensitive adenylate cyclase systems where several patterns of desensitization are observed. In S49 cells isoproterenol and PGE$_1$ produce specific desensitization[6] and although in human astrocytoma cells desensitization to these same agonists is initially specific, more prolonged exposure produces a component of non-specific desensitization[9].

TABLE II. Desensitization in guinea-pig ileum

Desensitizing agent	Test agonist			
	ACh	Hist	Sub. P.	K^+
Acetylcholine	+	+	+	−
Histamine	+	+	+	−
Substance P	−	−	+	−

+ Desensitization. − No desensitization.

Furthermore, since non-specific desensitization is not observed in membrane preparations it is clearly a process dependent upon cellular integrity. Ovarian adenylate cyclase exhibits both specific and partially non-specific desensitization since hCG produces desensitization to both LH and epinephrine, but not to PGE₁, whilst epinephrine produces specific desensitization only[11]. This latter system also demonstrates the previously noted differences between acute and chronic desensitization protocols since acute hCG treatment produces desensitization unaccompanied by true receptor loss whereas with chronic hCG treatment desensitization is accompanied by receptor loss. The existence of such systems, which show both specific and non-specific desensitization directed towards a common effector, is indicative of fundamental differences in mechanisms of adenylate cyclase activation and/or desensitization by catecholamines and glycoprotein hormones.

Similarly complex patterns of desensitization can be observed in other systems. Thus, guinea-pig ileum responds to a variety of smooth muscle excitants including acetylcholine, histamine, substance P etc. to generate mechanical response via an apparently common Ca^{2+} mobilization from the extracellular space. Both specific and non-specific desensitization can be observed in this system (Refs. 12–14; Table II). A major contributor to the non-specific component of desensitization is probably caused by membrane hyperpolarization caused by electrogenic pumping of the Na^+ ions that enter the smooth muscle cells during the stimulant-induced depolarization. This hyperpolarization will increase the threshold for excitation and the membrane will appear less sensitive. Consistent with this, Cantonini and Eastman[12] observed that elevated K^+ levels, which will prevent the hyperpolarization, reduced desensitization. However, the specific desensitization mediated by substance P (and some other agonists) must be caused by other events and conceivably represents changes at the level of the specific receptor.

Specific and non-specific desensitization is observed in immune responses. In IgE-mediated reactions desensitization to antigenic stimuli can be observed in systems from systemic anaphylaxis to *in vitro* mediator release. Desensitization of Ca^{2+}-dependent *in vitro* mediator release from basophils or mast cells can be achieved in several ways including challenge with high antigen concentration or exposure to antigen in the absence of Ca^{2+}. It has been suggested that the desensitization process involves the time-dependent closure of antigen-activated Ca^{2+} gates[15]. This Ca^{2+} activation process may involve bridging of IgE receptors but the mechanism leading to inactivation of the antigen-IgE stimulus is unknown. Of particular interest, however, is the influence of receptor density upon the desensitization process. Human basophils with low levels of bound IgE undergo specific desensitization to antigen challenge, which is not reversed by simple antigen removal, whereas basophils with high bound IgE undergo non-specific desensitization[16]. This suggests that specific desensitization may involve a specific change in the IgE molecule whereas non-specific desensitization is mediated only when a critical number of antigen-IgE antibody interactions occur. This latter process may involve the achievement of a threshold concentration of some intracellular messenger or the initiation of a global membrane perturbation.

Desensitization: functional aspects

The relationship of desensitization to cellular function is apparent in several ways. At the physiological level the loss of receptor sensitivity or actual receptor loss in acute and chronic desensitization respectively may serve to protect cells from excessive or prolonged stimulation, although presumably not from the initial stimulus. The receptor internalization which is a frequent accompaniment of chronic desensitization may serve not only as a mechanism for terminating the action of tightly bound hormone, but may also serve as an intracellular signal to regulate the rate of receptor biosynthesis. In any event, the process of receptor internalization in desensitization is a reflection of the equilibrium nature of cell membrane receptors, whereby variations in receptor density may be a normal component of cellular activity. This can be seen, for example, in the circadian variation of pineal gland β-adrenergic receptors[17], and in the desensitization of corpus luteum adenylate cyclase following ovulating doses of LH or hCG or mating (which causes an endogenous hCG surge) which may be a component of cellular signals initiating ovulation and luteolysis[18]. Different desensitization capacities of different cells may constitute a mechanism for hormonally mediated regulation of cell differentiation or function in a multicellular organism. Thus myeloid leukemic cells unable to desensitize to β-adrenergic agonists are vulnerable to the cytotoxic effects of the continued presence of these agonists[19].

In multiply sensitive systems specific desensitization may serve to permit sequential expression of cell signals. Thus in *Aplysia* ganglion cells which have acetylcholine receptors mediating depolarizing and hyperpolarizing responses with different kinetics of desensitization one or other response can be obtained according to stimulus frequency[20]. In contrast, nonspecific desensitization in multiply sensitive systems may serve to protect the system against all incoming stimuli.

At the pathologic level receptor loss has been associated with a number of disease states including myasthenia gravis, insulin resistance and allergic rhinitis where circulating receptor-directed antibodies increase degradation of membrane receptors. Receptor desensitization by excessive concentrations of the β-adrenergic stimulant has been associated with increased mortality in asthmatics.

It seems very probable that receptor-initiated desensitization should be regarded as one of the consequences of receptor activation and which can occur by several distinct mechanisms. The phenomenon of desensitization is often investigated under pharmacological conditions but it clearly also has physiological and pathological consequences. Continued research into this process will be of value not only for understanding desensitization, but also for strengthening our understanding of pharmacological receptors themselves.

Reading list

1 Catt, K. J., Harwood, J. P., Aquilera, G. and Dufau, M. L. (1979) *Nature (London)* 280, 109–116

2 Goldstein, J. L., Anderson, R. G. W. and Brown, M. S. (1979) *Nature (London)* 279, 679–685

3 Katz, B. and Thesleff, S. (1957) *J. Physiol. (London)* 138, 63–80

4 Heidmann, T. and Changeux, J. P. (1979) *Eur. J. Biochem.* 94, 255–279

5 Heidmann, T. and Changeux, J. P. (1978) *Annu. Rev. Biochem.* 47, 317–357

6 Shear, M., Insel, P. A., Melmon, K. L. and Coffino, P. (1976) *J. Biol. Chem.* 251, 7572–7576

7 Kent, R. S., De Lean, A. and Lefkowitz, R. J. (1980) *Mol. Pharmacol.* 17, 14–23

8 Mukherjee, C. and Lefkowitz, R. J. (1977) *Mol. Pharmacol.* 13, 291–303

9 Su, Y. F., Cubeddue, L. and Perkins, J. P. (1976) *J. Cyclic Nucleotide Res.* 2, 257–270

10 Nickols, G. A. and Brooker, G. (1979) *J. Cyclic Nucleotide Res.* 5, 435–447

11 Harwood, J. P., Defau, M. L. and Catt, K. J. (1979) *Mol. Pharmacol.* 15, 439–444

12 Cantonini, G. L. and Eastman, G. (1946) *J. Pharmacol. Exp. Ther.* 87, 392–399

13 Hurwitz, L. and McGuffee, L. J. (1979) in *Trends in Autonomic Pharmacol.* (Kalsner, S., ed.), Vol. 1, pp. 251–265, Urban and Schwarzenburg, Baltimore and Munich

14 Franco, R., Costa, M. and Furness, J. B. (1979) *Naunyn-Schmiedebergs Arch. Pharmakol.* 306, 195–201

15 Foreman, J. C. and Garland, L. G. (1974) *J. Physiol. (London)* 239, 381–391

16 Sobotka, A. K., Dembo, M., Goldstein, B. and Lichtenstein, L. M. (1979) *J. Immunol.* 122, 511–517

17 Kebabian, J. W., Zatz, M., Romero, J. A. and Axelrod, J. (1975) *Proc. Natl. Acad. Sci. U.S.A.* 72, 3735–3739

18 Hunzicker-Dunn, M. and Birnbaumer, L. (1976) *Endocrinology* 99, 211–222

19 Simantov, R. and Sachs, L. (1978) *Proc. Natl. Acad. Sci. U.S.A.* 75, 1805–1809

20 Gardner, D. and Kandel, E. R. (1977) *J. Neurophysiol.* 40, 333–348

The acetylcholine receptor

Jérôme Giraudat and Jean-Pierre Changeux

Institut Pasteur, Unité de Neurobiologie Moléculaire, 25, rue du Docteur Roux, 75724 Paris Cedex 15, France.

Neurons communicate with their target cells mostly by chemical means. The molecules involved, neurotransmitters, are in general small molecules which are liberated by nerve terminals upon rapid variation of the membrane potential. The change of neurotransmitter concentration in the synapse with time constitutes the 'signal' for inter-cellular communication.

At the neuromuscular junction in vertebrates or the electroplaque synapse in electric fishes[1,2], the physiological signal is a high concentration pulse of acetylcholine. In milliseconds the concentration in the cleft changes from 10^{-8} M to 10^{-4} M and back again and causes the all-or-none opening of ionic channels. These ionophores open only once for *ca.* 1–3 ms allowing small cations such as Na^+, K^+, and Ca^{2+} to traverse the postsynaptic membrane. At 20°C about 10^4 Na^+ ions flow through each ionophore into the cell. This rapid permeability response is triggered by the binding of acetylcholine to specific receptor sites present in the postsynaptic membrane. It is commonly referred to as *activation* and those compounds which give rise to the response as *agonists*. One of the most intriguing features of the activation is the sigmoid shape of the concentration effect curve (Hill coefficient of 1.5–3.0), and its striking analogy to the positive cooperative binding of ligands to regulatory proteins has often been pointed out.

When the agonist is applied for a prolonged period of time, rather than as a brief pulse, the amplitude of the permeability response slowly decreases, with a half-life of a few seconds. This *desensitization* of the membrane is reversed upon removal of the agonist.

The two responses of activation and desensitization are blocked by a group of pharmacological agents which interact in a competitive manner with acetylcholine. D-Tubocurarine and flaxedil are members of this group of *competitive antagonists,* as are the polypeptide α-toxins from snake venoms which, because of their high affinity (K_D *ca.* 10^{-10} M) and almost exclusive binding to the acetylcholine receptor site, have played a critical role in the identification of the protein that carries this site. Activation and desensitization are also sensitive to a heterogeneous group of compounds which do not bind to the acetylcholine receptor site at pharmacologically active concentrations. These compounds, which include the amine local anaesthetics and the frog toxin histrionicotoxin, decrease the amplitude of the permeability response and are therefore referred to as *noncompetitive blocking agents*[2]. The mechanism of their action is not yet clearly understood. They enhance desensitization and may as a consequence decrease the number of receptor molecules available for activation; on the other hand, because they modify the shape of the endplate potential and of the single channel current pulses these compounds have been postulated to bind directly to the open ion gate and to act there as steric analogues of the permeant ion. Structural analogies may exist between some of the ligands binding to the acetylcholine receptor site and typical non-competitive blockers eventually leading to mixed effects. In any case, the non-competitive blockers, exclusive or not, are useful tools for identifying the structures involved in the translocation of ions and their coupling with the acetylcholine receptor site.

Model of the acetylcholine regulator

Acetylcholine acts as a regulatory ligand which, by the simple virtue of binding to its specific receptor site, modifies the state of opening of a nearby and closely linked ion channel. This is why it was postulated that the elementary membrane unit, or acetylcholine *regulator,* carrying the receptor site for acetylcholine and the site for ion translocation is made up of at least two distinct structural elements: the receptive unit, or acetylcholine *receptor* (AChR), and the biologically active unit, or *ionophore,* which might be part of the same polypeptide chain or equally be carried by distinct subunits.

By analogy with the known behaviour of regulatory enzymes and in agreement with the Katz and Thesleff model for desensitization, it was further postulated[2,3] that the ACh-regulator may exist in *at least* three discrete and interconvertible conformational states. These states, referred to as resting R, active A and desensitized D, would differ primarily in the affinity of the AChR site for cholinergic ligands. The affinity for agonists would increase from R (low affinity) to A (medium affinity) to D (high affinity); on the other hand, the

affinity for competitive antagonists would be lower in the A than in the R or D states. Postulated conformational states of the acetylcholine regulator may also differ in their affinity for non-competitive blocking agents; in the D and A states the affinity for these compounds would be higher than in the R state. Finally, the ion gate would be open only in the A state of the regulator. Of course, this is a minimal model; more states might have to be involved to account for the experimental data, and cooperative interactions between acetylcholine binding sites might have to be added.

Another important feature of this model, which accounts for the characteristics of the physiological acetylcholine 'signal' at the synapse, is that in the membrane at rest the regulator would be present in the R state, which has a low affinity for acetylcholine, whereas at equilibrium acetylcholine would stabilize the high affinity desensitized D state. Upon rapid change of ligand concentration and depending on the nature of this ligand, some states can be transiently populated owing to preferential pathways for the interconversion. The 'activation' reaction is viewed as a transient population of the A state by acetylcholine and the 'desensitization' as the final equilibration in the D state.

Structure of the subsynaptic membrane

As soon as they were introduced as selective labels of the acetylcholine (nicotinic) receptor site, the snake venom α-toxins were employed to study, first qualitatively then quantitatively, the distribution of the AChR at the neuromuscular junction or the electroplaque synapse[2]. The observed high surface densities (up to 30,000 toxin sites per μm^2 in certain subsynaptic areas) can be explained only if the subsynaptic membrane contains almost exclusively receptor–ionophore complexes cemented with lipids. The course of the studies on the AChR shifted dramatically when suspensions of such subsynaptic membrane fragments became available. This was achieved by the differential centrifugation of homogenates of the electric organ from *Torpedo,* a tissue exceptionally rich in cholinergic synapses. In these fragments, almost depleted of acetylcholinesterase, the receptor protein may represent up to 40% of the total and *ca.* 140 mg of such fragments can be obtained from 1 kg of electric tissue[4]. They make closed microsacs with which $^{22}Na^+$ efflux or influx can be measured by a filtration assay[5]. In the presence of agonist the rate of passive $^{22}Na^+$ (and K^+ or Ca^{2+}) transport increases and this response is blocked by the same series of competitive and non-competitive effectors as in the live electroplaque or neuromuscular junction. Moreover, prolonged exposure of the membrane fragments to agonists causes a time-dependent decrease of response amplitude: desensitization is still present. In other words, with these preparations a wide spectrum of physicochemical methods may be used to study the functional receptor protein in its membrane environment.

When examined by electron microscopy after negative staining or deep etching, these receptor-rich membranes disclose transmembrane particles which' have the same size and shape as the characteristic rosettes observed in preparations of purified AChR protein: a diameter of 8–9 nm, a central pit *ca.* 1.5 nm wide, and 5–6 subunits[6]. These particles are present on the etched surfaces and are therefore exposed on the outside of the membrane fragments. Similar images have been recently reported on the cleft surface of the subsynaptic membrane *in situ* after rapid freezing[7]. The surface density of these rosettes compares well with that of the α-toxin sites if one assumes several toxin sites (about 3) per rosette.

Clusters of densely-packed receptor rosettes may display regular arrangements into double lines or pseudohexagonal arrays. Also, X-ray diffraction studies of membrane fragments oriented by centrifugation reveal a regular organization of particles within the plane of the membrane; they also disclose that dense material may protrude up to 55 Å from the classical lipid bilayer (see Ref. 7). On the other hand, 'internal' particles that

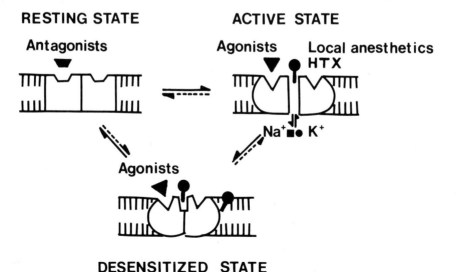

RESTING STATE   **ACTIVE STATE**

DESENSITIZED STATE

Fig. 1. A model of the acetylcholine regulator (see text).

follow the distribution of the receptor-rosettes are observed in the fracture plane but their significance is not yet clear.

The protein-to-lipid ratio of these purified membrane fragments might be as high as 1.5–2.3 (w/w). Cholesterol is a dominant component of the lipids (weight ratio to phospholipids 0.40) and long chain phospholipids are particularly abundant. Electron paramagnetic resonance spectroscopy of the receptor protein spin-labelled by a covalent probe shows an almost complete immobilization of the receptor (rotational correlation time > 1 ms as against 20 μs at 20°C for rhodopsin in fluid rod outer segment membrane) presumably resulting from protein–protein interactions within the subsynaptic membranes[8].

The analysis of the polypeptide chain composition of the receptor-rich membrane fragments has been performed by polyacrylamide gel electrophoresis in one dimension in the presence of the denaturing detergent sodium dodecyl sulphate. Different results were reported for T. californica and T. marmorata, E. electricus and skeletal muscle receptor[4,9-11]. However, the authors agree today that the variability resulted from a proteolytic nicking of some of the polypeptide chains. Two major components have an apparent mol. wt. of 40 000 and 43 000. Besides these, chains of apparent mol. wt. of 50 000, 60 000 and 66 000 are present. The accepted molar ratio of the 40 000, 50 000, 60 000 and 66 000–Mr polypeptides is 2-1-1-1[9,10,22].

One function of the 40 000 Mr polypeptide is clear. This chain is the only one labelled by affinity reagents specific to the AChR site and this labelling is completely destroyed by preincubation with snake α-toxins[10,12]. Without ambiguity the 40 000 Mr polypeptide carries the physiological receptor site for acetylcholine.

Photoaffinity labelling experiments, carried out with various organic compounds[9,10], and with derivatives of snake α-toxins, give an incorporation of radioactivity into the 50 000 and 66 000 Mr chains in addition to the 40 000 Mr one, suggesting that the 50 000 and 66 000 Mr chains lie in the close vicinity of the 40 000 Mr one[10].

The 43 000–Mr polypeptide is present in the native membrane in approximately equal amounts to the 40 000 subunit. It can be separated from the native receptor-rich membrane by treatment at pH 11 without complete disruption of the membrane structure[13]. It appears therefore to be a peripheral rather than an integral membrane protein. It has been claimed that it represents actin. Actin is indeed present in the electric organ but does not co-migrate with nor have the same amino acid composition as the 43 000 Mr polypeptide.

The transmembrane organization of these various polypeptides has been studied by selective proteolysis[14] due to the fact that the receptor-rich membranes from T. marmorata spontaneously reseal with the right side out. The 40 000, 50 000, and 66 000 Mr polypeptides are attacked by proteases both from inside (after opening of the membranes, e.g. by ultrasound) and from outside (after exposure to acid pH). All of them, therefore, traverse the membrane. The same conclusion has been subsequently extended to T. californica receptor. On the other hand the 43 000 Mr chain is degraded by proteases exclusively from the inside of the microsacs and is therefore probably bound to the inner face of the subsynaptic membrane[14].

Properties of the purified receptor protein

The AChR, being an integral component of the subsynaptic membrane, cannot be solubilized without detergents. Non-denaturing detergents disperse the receptor-rich membranes from Torpedo into a minimum of two macromolecular entities: one made up exclusively of the 43 000 Mr polypeptide and another which contains the 40 000, 50 000, 60 000 and 66 000-Mr chains. This last species, which still binds the snake venom α-toxins, covalent affinity reagents of the receptor site and various cholinergic ligands, is referred to as the acetycholine *receptor protein*. It can be purified to homogeneity in mg quantities by ultra-centrifugation of a crude detergent extract of *Torpedo* receptor-rich membranes[4] but, originally and still for *E. electricus,* affinity chromatography using cholinergic ligands covalently coupled to a solid matrix has been most efficiently used for purification. The purest fractions of receptor protein bind

8–12 μmol of α-toxin per g protein[2] (mol. wt. per site *ca.* 100 000). When injected into various experimental animals, the receptor protein from fish causes an immune reaction which resembles the human disease *myasthenia gravis*[15]. The antisera from these animals contain antibodies which block the *in vivo* response of the electroplaque to acetylcholine confirming the physiological significance of the purified protein.

Polypeptide compositions reported for the receptor protein extracted from *Torpedo* receptor-rich membranes resemble those of the native membranes. That is 2×40: 1×50: 1×60: 1×66 kilodaltons polypeptides. In *Electrophorus,* two minor

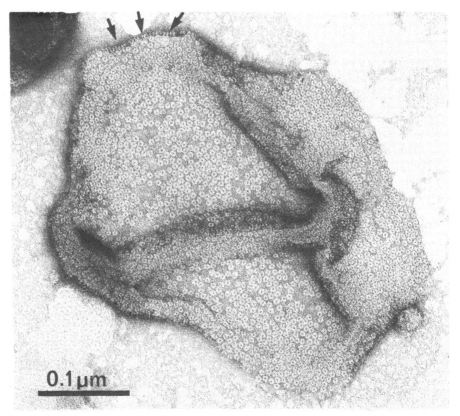

Fig. 2. Electron micrograph of acetylcholine receptor-rich membrane fragment from Torpedo marmorata *(negative staining × 250 000). Made by Jean Cartaud.*

The surface of the membrane is covered with rosettes which represent the receptor molecules. Arrows in the upper left corner point to the edge of the vesicle where double projections are visible. These projections, protruding from the membrane surface are suggestive of the exposure of a substantial part of the molecules at the external surface of the membrane.

polypeptides of apparent mol. wt. 48 000 and 54 000 have been reported in the past[10] in addition to the dominant 40 000 Mr one. However, this simplified pattern results, again, from the proteolytic cleavage of a molecule having the same structure as *Torpedo* receptor. This is also the case for muscle receptor. Antibodies raised against any of the polypeptide chains precipitate the whole receptor protein and give rise to a myasthenia like reaction in rats[15].

Sedimentation in sucrose gradients of crude detergent extracts of purified receptor protein from *Torpedo* reveals two major species: a light (L) form, of standard sedimentation coefficient 8–9 S, and a heavy (H) form of 12–13 S. Pre-incubation with mercaptoethanol or dithiothreitol causes an almost complete conversion of H into L. Disulphide bonds are thus likely to be involved in the L–H transition. Denaturing gel electrophoresis reveals no 66 000 Mr band in the H form but a 130 000 Mr band which is transformed into the 66 000 Mr one concomitantly with the conversion of H into L. This suggests that the H form results from an intermolecular disulphide bond cross-linking two 66 000 Mr polypeptides from separate L molecules. The reported molecular weight of the L form ranges between 250 000 and 275 000 depending on the technique used[2] and the mol. wt. of H is consistent with it being a dimer of L.

The purified receptor is a rather acidic (pHi 4.8–5.3) and hydrophobic protein (it may bind up to 20% of its weight in detergent). Each of its constituent polypeptide chains carries a carbohydrate moiety which binds plant lectins and contains mannose, glucose and galactose residues. Interestingly, the 50 000 and 66 000 Mr polypeptides react with phytohaemagglutinin A but not the 40 000 Mr one[14]. The 40 000 Mr polypeptide contains *O*-substituted serines and threonines among them phosphoserines

which might possibly be relevant to the phosphorylation observed *in vitro*. Its amino acid composition reveals a significant hydrophobic character close to that of rhodopsin. The NH_2-terminal sequence of 20 amino acids has been determined[16] as follows:

1	5	10

Ser – Glu – His – Glu – Thr – Arg – Leu – Val – Ala – Asn –

	15	20

– Leu – Leu – Glu – Asn – Tyr – Asn – Lys – Val – Ile – Arg.

Ligand binding sites and allosteric transitions of the acetylcholine regulator

The physiological receptor site for acetylcholine is easily identified by its ability to bind *reversibly* cholinergic agonists, competitive antagonists and snake α-toxins. All of them bind with the same one to one stoichiometry[23] and exhibit competitive interactions with each other. A structural analogy has been pointed out between a discrete area of the X-ray diffraction model of an α-toxin and quaternary cholinergic ligands. Most likely these ligands bind to overlapping areas on the receptor surface. An apparent heterogeneity of binding has been noticed for α-toxin and D-tubocurarine (see Refs. 2 and 23 as if negative cooperative interactions were taking place between their respective sites on the resting receptor molecule. Positive cooperative binding has also been observed between acetylcholine binding sites[2].

Spectroscopic experiments and direct binding studies with fluorescent or radioactive local anaesthetics and histrionicotoxin have confirmed that the noncompetitive blockers bind to *Torpedo* receptor-rich membranes at a population of binding sites that can be saturated and are distinct from the AChR sites[2,17]. Interestingly, reciprocal positive interactions are established between these two classes of sites: at equilibrium, local anaesthetics, in a domain of concentration where they block ion transport, enhance acetylcholine binding[2,3]; agonists exert the same effect on high affinity binding of the non-competitive blockers[17]. Ratios of 0.3–0.75 local anaesthetic sites per α-toxin or acetylcholine binding site have been reported, these being located not on the 43 000 Mr chain but on the 66 000 Mr polypeptide[22]. Their precise relation to the acetylcholine ionophore and to the 40 000 Mr peptide which carries the AChR site is not yet understood.

The equilibrium binding properties of the receptor protein in its purified and

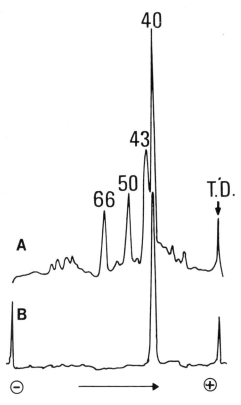

Fig. 3. Polyacrylamide gel electrophoresis in the presence of sodium dodecyl sulphate of acetylcholine receptor-rich membrane fragments from Torpedo marmorata. Membrane fragments have been labelled with an affinity reagent specific to the acetylcholine receptor site: [³H]MPTA[10]. (A) Protein bands on a 10% acrylamide gel. T.D., tracking dye. (B) Fluorography of the gel.

membrane-bound forms remained a confused issue for years, (for instance, under equilibrium conditions, acetylcholine binds to its specific site on *Torpedo* receptor-rich membranes with a dissociation constant of 10 nM which is *ca.* 3–4 orders of magnitude below the concentration at which it is active on membrane permeability) until both theoretical (see model) and experimental[3] evidence was proposed for the existence of several interconvertible states of the receptor protein. Under equilibrium conditions, acetylcholine would stabilize a high affinity 'desensitized' form of the receptor and its 'active' conformation would not be accessible by equilibrium measurements but by dynamic binding techniques. This proposal was rapidly confirmed, in particular by quantitative measurement of the interaction of fluorescent[18] or radioactive (Cohen and Boyd, in press) ligands in the ms–s time scale using rapid mixing techniques. The fluorescent agonist Dns–C₆–Cho which behaves like acetylcholine on the electroplaque gives, when it interacts with the acetylcholine receptor site, a fluorescent signal which can be recorded under conditions of energy transfer from proteins. A quantitative analysis of the stopped-flow traces at various concentrations of Dns–C₆–Cho reveals a minimum of three relaxation processes: (1) a 'rapid' process takes place in the ms time range with an amplitude that corresponds to approximately 20% of the total signal and is interpreted as the result of Dns–C₆–Cho binding to a population of AChR-sites with a *high* affinity for Dns–C₆–Cho (K_D *ca.* 3 nM) which represents approximately 20% of the total population of receptor-sites and exists in the membrane at rest prior to agonist binding; (2) an 'intermediate' relaxation process is recorded in the s–ms time range. It becomes important at high (10^{-4} M) Dns–C₆–Cho concentration and is accounted for, at a first approximation, by a bimolecular binding reaction to a lower affinity state (or states) of the receptor-site; (3) a 'slow' relaxation process taking place in the s time range is interpreted as representing an isomerization between the above mentioned low and high affinity states of the receptor-protein.

The experimental data are consistent at a first approximation with a simplified version of the three-state model presented earlier. Binding to the low affinity state(s) would include the 'activation' process and the isomerization to the high affinity state would correspond to the 'desensitization' reaction[19]. This last interpretation is supported by the observation that the non-competitive blockers which accelerate desensitization *in vivo* also increase the apparent rate constant of the slow transition leading to this high affinity state[18]. Local anaesthetics also accelerate the 'intermediate' relaxation process suggesting that it includes an isomerization towards a third state to which the local anaesthetics would also bind. This still unresolved transition would correspond to the 'activation' reaction. It might be identical to the transition revealed in the presence of the fluorescent local anaesthetic quinacrine.

Reconstitution of a functional acetylcholine regulator

A particularly suitable and elegant method for identifying the components of the subsynaptic membrane that are necessary and sufficient for the physiological response is to *reconstitute* an excitable membrane from chemically defined components in solution. In the first attempts, the regulation of $^{22}Na^+$ flux was selected as the principal criterion for the 'functional' state of the receptor protein. The experiment appeared feasible after elimination of the detergent (by dialysis, filtration on gel, etc.) from a soluble extract of purified receptor-rich

membranes[20] or even after purification of the receptor protein (see Ref. 21). However, measurement of agonist-sensitive ion fluxes is not sufficient to establish a complete recovery of the native properties of the acetylcholine regulator. In a reconstituted vesicle, a few 'functional' molecules among a large fraction of inactive ones may indeed permit significant ion fluxes. A quantitative estimation of the fraction of receptor molecules in an 'active' state was made possible by following throughout the reconstitution the allosteric transition of the receptor protein with the fluorescent agonist Dns–C_6–Cho. In addition to the recovery of agonist-sensitive ion fluxes, the presence of a significant fraction of receptor sites in a low affinity state, their slow interconversion to the high affinity state by agonists and the effects of non-competitive blockers on this interconversion were therefore considered as essential criteria for reconstitution. Recent experiments[21] indeed show that conditions of reconstitution might be defined where, starting from membrane fragments, the end product satisfies all these requirements.

Since reconstitution can be carried out under conditions where 95% of the receptor protein is in its 9 S form it is concluded that this species is indeed functional (rather than exclusively the 12 S dimer). No difference was noticed when alkali-treated membranes were used instead of native membranes as starting material for the reconstitution, confirming that the 43 000 Mr protein does not play a significant role in the allosteric properties of the receptor protein nor in the process of reconstitution itself. Only the 40 000 Mr chain and the other three receptor polypeptides presently appear critical.

The major finding of these reconstitution experiments concerns the critical role played by lipids. Reconstitution is reproducibly achieved as long as the total concentration of lipids (which might be exogenous, such as a crude soybean lipid extract) remains elevated in the soluble extract. In fact, under these conditions, the receptor protein *in solution* (9 S) still shows most of the characteristic binding properties of the membrane-bound receptor: low affinity, interconvertibility to high affinity, and effect of local anaesthetics[21,24].

In this respect it should be recalled that the receptor protein is an integral protein which, as such, might be highly sensitive to the physical state and biochemical composition of membrane lipids. Several compounds known to interact with the lipid phase of the membrane indeed block the permeability response to acetylcholine both *in vivo* and *in vitro* and even modify the binding properties of the membrane-bound acetylcholine receptor. Also, a direct interaction between the purified receptor protein and lipid monolayers has been demonstrated.

In conclusion, the recent investigations from different laboratories on the acetylcholine receptor protein from fish electric organs have led to a precise definition and understanding of this molecule in chemical and functional terms although several questions as to the nature of the receptor polypeptide carrying the ion gate remain unanswered. In many regards the receptor protein from the neuromuscular junction of higher vertebrates appears similar. Studies in progress on receptors for other neuro-transmitters and from other tissue, such as brain, should tell us to what extent all ionophore-coupled receptors have proper-ties analogous to those of the acetylcho-line receptor from the fish electric organ.

Reading list

1 Colquhoun, D. (1979) in *The Receptors General Principles and Procedures,* Vol. 1, pp. 93–142 (O'Brien, R. D., ed.) Plenum Press

2 Heidmann, T., and Changeux, J. P. (1978) *Ann. Rev. Biochem.* 47, 371–441

3 Changeux, J. P., Benedetti, L., Bourgeois, J. P., Brisson, A., Cartaud, J., Devaux, P., Grünhagen, H., Moreau, M., Popot, J.L., Sobel, A. and Weber, M. (1976) in *The Synapse, Cold Spring Harbor Symp. on Quantitative Biology,* Vol. XL, pp. 211–230

4 Sobel, A., Weber, M. and Changeux, J. P. (1977) *Eur. J. Biochem.* 80, 215–224

5 Kasaï, M. and Changeux, J. P. (1971) *J. Membr. Biol.* 6, 1–80

6 Cartaud, J., Benedetti, L., Sobel, A. and Changeux, J. P. (1978) *J. Cell. Sci.* 29, 313–337

7 Heuser, J. E. and Salpeter, S. R. (1979) *J. Cell. Biol.* 82, 150–173

8 Rousselet, A., Devaux, P. F. and Wirtz, K. W. (1979) *Biochem. Biophys. Res. Commun.* 90, 871–877

9 Raftery, M., Blanchard, S., Elliott, J., Hartig, P., Moore, H., Quast, U., Schimerlik, M., Witzemann, V. and Wu, W. (1979) in *Advances in Cytopharmacology, Neurotoxins: Tools in Neurobiology* Vol. 3, pp. 159–182 (Ceccarelli, B. and Clementi, F., eds.) Raven Press, New York

10 Karlin, A., Damle, V., Hamilton, S., McLaughlin, M., Valderamma, R. and Wise, D. (1979) in *Advances in Cytopharmacology, Neurotoxins: Tools in Neurobiology* Vol. 3, pp. 183–189 (Ceccarelli, B. and Clementi, F., eds.) Raven Press, New York

11 Hucho, F., Layer, P., Kiefer, H. and Bandini, G. (1976) *Proc. Natl. Acad. Sci. U.S.A.* 73, 2624–2628

12 Weiland, G. and Taylor, P. (1979) *Mol. Pharmacol.* 15, 197–212

13 Neubig, R. R., Krodel, E. K., Boyd, N. D. and Cohen, J. B. (1979) *Proc. Natl. Acad. Sci. U.S.A.* 76, 690–694

14 Wennogle, L., and Changeux, J. P. (1980). *Eur. J. Biochem.* 106, 381–393

15 Lindström, J. (1979) in *Advances in Cytopharmacology, Neurotoxins: Tools in Neurobiology* Vol. 3, pp. 245–253 (Ceccarelli, B. and Clementi, F., eds.) Raven Press, New York

16 Devillers-Thiéry, A., Changeux, J. P., Paroutaud, P. and Strosberg, A. D. (1979) *FEBS Lett.* 104, 99–105

17 Krodel, E. K., Beckman, R. A. and Cohen, J. B. (1979) *Mol. Pharmacol.* 15, 294–312

18 Heidmann, T. and Changeux, J. P. (1979) *Eur. J. Biochem.* 94, 255–279 and 281–296

19 Sine, S. and Taylor, P. (1979) *J. Biol. Chem.* 254, 3315–3325

20 Hazelbauer, G. L. and Changeux, J. P. (1974) *Proc. Natl. Acad. Sci. U.S.A.* 71, 1479–1483

21 Changeux, J. P., Heidmann, T., Popot, J. L. and Sobel, A. (1979) *FEBS Lett.* 105, 181–187

22 Saitoh, T., Oswald, R., Wennogle, L., and Changeux, J. P. (1980). *FEBS Lett.* 116, 30–36

23 Neubig, R., and Cohen, J. B. (1979). *Biochemistry* 18, 5464–5475

24 Heidmann, T., Sobel, A., Popot, J. L., and Changeux, J. P. (1980). *Eur. J. Biochem.* 110, 35–55

Early work on the adrenergic mediator: how to go wrong

Z. M. Bacq

Laboratoire de Biochimie Appliquée, 32, Bd de la Constitution, 4020—Liège, Belgium.

The usefulness of error

I have been told by a trustworthy friend that Maurice Ravel - the famous French composer well known for his ability to manipulate the multiple instruments of the modern orchestra - asked his colleague Georges Auric to write with him a *treatise of orchestration* illustrated by a lot of examples, showing what one *has to avoid* instead of the usual *good* models. Indeed, error is often more instructive than success when you understand its genesis and can put your finger on the wrong step.

It happens that I have been involved since 1929 - just half a century ago - in work on the chemical mediation of adrenergic nerve stimulation and that I have first-hand information and can follow step by step the progress of the facts and of their interpretation in this field[1].

W. B. Cannon, Harvard 1929

From September 1929 to December 1930, I was research fellow in the Department of Physiology of the Harvard Medical School headed by W. B. Cannon who was at the acme of his fame[2]. He had just presided over the 13th International Congress of Physiology attended by all the prestigious scientists of that time including I. P. Pavlov and Léon Fredericq. W. B. Cannon had performed preliminary experiments showing the possibility of release into the blood of a sympathomimetic substance when sympathetic nerves to the liver were stimulated *in the intact cat*; he wanted to confirm these observations. I became his personal assistant and was in charge of the lengthy aseptic preparation of the cats; surgical denervation of the heart, removal of the right adrenal and denervation of the other. The acute experiment was done ten days after when the heart was sensitized.

In 1930, when W. B. Cannon came back from a tour of Europe, his health was not excellent and I tried to relieve him from the greatest part of the material load. All the final experiments were done by both of us. It appeared clearly that, in the intact cat, electrical stimulation of sympathetic nerves (lumbar chain or splanchnic) induced a delayed acceleration of the denervated heart due to the liberation of a sympathomimetic substance in the circulating blood. We found that ergotamine *increased* this acceleration, a phenomenon rediscovered in 1947 by G. L. Brown and J. S. Gillespie and correctly interpreted many years after by the presence of presynaptic α-receptors. Since we had no idea about the chemical nature of this substance, we proposed to call it 'sympathin' in the paper published in 1931 in the *Am. J. Physiol.*[4]. Naturally, we had in mind that sympathin was adrenaline itself, as suggested by Elliott in

1904. At that time W. B. Cannon was aware of the fact that the actions of the natural adrenaline extracted from adrenal glands were not exactly the same as that of the synthetic substance; he used the term *adrenin*. He was right; all the samples of natural adrenaline available at that time were contaminated with a small proportion of noradrenaline.

Two sympathins

After my return to Belgium, Dr Cannon continued working in the same field with Arturo Rosenblueth, who had arrived at Harvard in October 1930 and had witnessed the final rather spectacular experiments on sympathin. The tests for sympathin were changed to the nictitating membrane and the non-pregnant uterus; the first muscle is excited by catecholamines, the second is inhibited. The cats were anesthetized with a barbiturate. Cocaine* was used to sensitize the smooth muscles. They stimulated the peripheral end of the cut autonomic fibers along the hepatic artery. When the effects of liver sympathin were compared with those of intravenously injected epinephrine, it appeared that sympathin was less effective in the tests of inhibitory action. It seemed that if that part of the hepatic nerve which innervates the duodenum (which was supposed to have inhibitory effects) were severed or excluded from the electric stimulation, the small inhibitory effect on the uterus was decreased. But, to be sure, it was not possible to obtain pure inhibitory or pure excitatory effects with any kind of sympathin.

The main and correct observation of Cannon and Rosenblueth[5] was that, as compared with adrenaline, the action of sympathin on the non-pregnant cat uterus (inhibition, now recognized as a β-effect) was smaller than that on the nictitating

membrane (contraction, a typical α-effect). The interpretation of this fact was very curious indeed; it was supposed that the chemical mediator M liberated by the excitation of the nerve ending reacted with a receptor E (for the excitatory effect) or I (for the inhibitory effect) *and that it was the combination* (ME = sympathin E purely excitatory; MI = sympathin I purely inhibitory) *which diffused in the blood.* Thus, if one reads this paper of Cannon and Rosenblueth carefully, it is obvious that there is to be found the first hint of two different types of receptors which R. P. Ahlquist logically coined α and β in 1948. But the unforgivable, awful error, lacking in factual or theoretical arguments, was to postulate that the combination of the mediator with the receptor could be separated from the membrane of the effector cell and could move in the circulating fluids. Nothing in Langley's papers suggested such a possibility.

Who is the culprit?

In my opinion, there is no question that this unfortunate idea of the two sympathins originated in Rosenblueth's mind†.

†One finds in Cannon and Rosenblueth's paper[5] the following sentences, "An analysis of the action of adrenin and of nerve impulses on smooth muscle led Rosenblueth to the hypothesis (consistent with his experimental results) that a substance A from the outside (e.g. adrenin) or M (a local product dependent on the number of nerve impulses in the stimulus) unites in the cell with another substance H, thus making a combination AH or MH which evokes a response proportional to the amount formed. As records of the intestine and the non-pregnant uterus prove, the conception is quite as applicable to inhibition as it is to excitation; consequently H must be regarded as either I (inhibitory) or E (excitatory) and the combination, after nerve stimulation, for example, would be ME in a contracting muscle and MI in a relaxing muscle. Sympathin is defined as the chemical mediator of sympathetic nerve impulses, ME or MI, which in the cell induces the typical response, contraction or relaxation, and which, escaping from the cell into the blood stream, induces effects elsewhere in organs innervated by the sympathetic.

*A. Fröhlich and O. Loewi had discovered this action of cocaine in 1910.

Rosenblueth was a very intelligent, brilliant, hard-working fellow, fond of speculations and mathematical analysis as shown by his many papers published in the years 1932–1950.

I went back to Harvard in 1933 and worked for a couple of months with Rosenblueth, but failed to persuade him that his hypothesis was not only improbable but useless. Sad to say, I became a kind of heretic not too kindly attacked in various papers and books. Cannon attended the 15th International Congress of Physiology (Moscow–Leningrad) in 1935. It was the last time I saw him. A. Rosenblueth was not present. There was very little talk about two sympathins, but W. B. Cannon delivered a courageous address which precipitated a walkout of the German Nazi delegation. Even in 1950, five years after W. B. Cannon's death, Rosenblueth was still desperately defending the two sympathins hypothesis in a forgotten lengthy book[8] despite the accumulation of evidence by U. S. von Euler in favor of noradrenaline as the main component of catecholamines released by adrenergic nerve stimulation in mammals.

The European side (1931–1934)

I was very sorry indeed. W. B. Cannon inspired in his associates both affection and admiration. He was a true, generous democrat; he spent a lot of time in 1937–1939 campaigning in the U.S.A. for the Spanish republicans. But I could not avoid standing clearly against the two sympathins theory. I said so in a report to the Association des Physiologistes in Nancy, 1934. This paper written in French and published in the *Annales de Physiologie* (which disappeared in 1940) did not pass the language barrier and was completely ignored. I remember that in 1933, I met M. Tiffeneau at the Institut Pasteur in Paris in the office of E.

Fourneau whose chemists had just prepared the very first synthetic α-antagonists; the most popular of this series is the F. 933 (piperidinomethy-benzodioxane) known at present as piperoxan. They were looking for antimalarials; Daniel Bovet by chance discovered that they inversed the pressor effect of adrenaline in the dog very much like ergotamine. Not a single β-antagonist was known at that time. Tiffeneau repeated that he could not understand why 'Nature' had apparently failed to use 'arterenol' (= noradrenaline) a more powerful vasoconstrictor than adrenaline and, chemically speaking, a simpler substance. I had been much impressed by this fact and read again the famous paper of Barger and Dale[3] on the relation of chemical structure and actions of amines.

Noradrenaline (which appears in Barger and Dale's paper as amino-ethanol-catechol) was already known in 1910. It had been tested by Loewi and Meyer in 1905 (*Arch. f. Exp. Path. v. Pharm.* 53, 213) in connexion with the synthetic work of Stolz (*Ber. deutsch. Chem. Ges.* 1904, 37, 4149) and Dakin (*Proc. R. Soc.* 1905, 76B, 491). It was commercially available in 1910 from the Hoechst Company and Barger was not obliged to prepare it. Sir Henry mentions that when he wrote this famous paper with G. Barger, an elaborate physiological comparison between adrenaline and arterenol by Schultz (*Hygenic Laboratory Bulletin,* No. 55, Washington 1909) had already been published.

I pointed out in my 1934 report that noradrenaline had all the properties of sympathin E and that adrenaline acted more like sympathin I, the existence of which remained doubtful. Even Cannon and Rosenblueth[5] admitted this difficulty; "The simple, positive evidence of the existence of sympathin I, apart from E, we have not obtained." But, unfortunately at that time the chemical techniques were so primitive that it was unthinkable to separate noradrenaline from adrenaline. Paper chromatography did not exist; it took about 4 h to obtain a poor u.v. absorption spectrum of a couple of solutions of frog's heart perfusate. The conclusion of my 'thèse d'agrégation' (*Arch. Intern. Physiol.* 1933, 36, 167) was that, taking into consideration all that was known at that time about its physiological effects and physicochemical properties, sympathin must be a derivative of catechol: ethylamino, ethanolamino or methylethanolamino. There was no objection to suggesting that it might be a mixture of amines.

Sir Henry H. Dale's verdict

When in 1952, Sir Henry was given the opportunity to reprint his main papers he added most interesting comments on his big paper (1910) with G. Barger on the action of amines[6]. In 1910, Sir Henry was not in favor of Elliott's hypothesis that in the sympathetic system adrenin is the chemical stimulant liberated whenever the impulse arrives at the periphery; he thought that it introduced unnecessary complications. But, T. R. Elliott was (and remained) one of Sir Henry's best friends; indeed the "Adventures in physiology with excursions into autopharmacology" were dedicated to him. Sir Henry took great pains to discuss Elliott's suggestion and said clearly that primary amines would be better candidates than adrenin for the function of a chemical link between the nerve ending and the cell membrane. In his own words: "Doubtless, I ought to have seen that noradrenaline might be the main transmitter – that Elliott's theory might be right in principle and faulty only in this detail. If I had so much insight, I might even then have stimulated my chemical colleagues to look for *nor*-adrenaline in the body . . .But if I had taken the additional step, even in hypothesis, much trouble might, perhaps,

have been saved in after years for my late friend, Walter Cannon. For the observed differences between the effects of adrenaline and those of 'sympathin' as liberated by stimulation of sympathetic nerves, and especially of the hepatic nerves, which led Cannon and his associates to put forward the elaborate theory of the two complex sympathins, E. and I, were practically the same as those between adrenaline and nor-adrenaline, with which I was here concerned. It should be noted, indeed, that one who had collaborated with Professor Cannon at one stage, Professor Z. M. Bacq, actually suggested in 1934 that 'Sympathin E' might be *nor*-adrenaline *(Ann. d. Physiol.* 1934, 10, 480); but, in the absence of direct evidence then, and for many years afterwards, of the natural occurrence of *nor*-adrenaline in the animal body, this suggestion attracted less attention than it deserved".

What is the present situation?

The substances liberated by nerve impulses or potassium ions and stored in the specific granules at the adrenergic nerve ending are mixtures in variable proportions of noradrenaline, its precursor dopamine and its N-methylated derivative. Living organisms never make 100% pure substances.

O. Loewi was right, by pure luck: his 'Acceleransstoff' is mainly adrenaline. The Amphibians are powerful N-methylators. In the organs of fishes, von Euler (1953) found about equal amounts of adrenaline and noradrenaline. In Birds, there is a large proportion (20–30%) of adrenaline. In Mammals, the dominance of noradrenaline is such that the small amounts of dopamine and adrenaline (for instance in the blood perfused through the stimulated cat spleen) may be considered as negligible. The Cephalopods, the most highly organized Molluscs, synthesize a splendid variety of mono- and diphenolic amines besides histamine and 5-hydroxytryptamine; but for the main quantity, they prefer to play with the monophenolic ones. The comparative pharmacology of amines is a marvellous field full of unexpected observations which Hans Fischer has meticulously organized in an exhaustive monograph[7].

Reading list

1. Bacq, Z. M. (1976) *J. Physiol. (Paris)* 72, 371–473.
2. Bacq, Z. M. (1975) In: Brooks Chandler, McC., Koizumi, K. and Pinkston, J. O. (eds), *The life and contributions of Walter Bradford Cannon,* Downstate Medical Center, New York, pp. 68–83.
3. Barger, G. and Dale, H. H. (1910) *J. Physiol. (London)* 41, 19–59.
4. Cannon, W. B. and Bacq, Z. M. (1931) *Am. J. Physiol.* 96, 392–412.
5. Cannon, W. B. and Rosenblueth, A. (1933) *Am. J. Physiol.* 104, 557–574.
6. Dale, H. H. (1953) *Adventures in Physiology,* Pergamon Press, Oxford.
7. Fischer, H. (1971) *Vergleichende Pharmakologie von Uberträgersubstanzen in tiersystematischer Darstellung; Handbook of Exp. Pharm.,* Vol. XXVI, Springer, Berlin.
8. Rosenblueth, A. (1950) *The Transmission of Nerve Impulses at Neuroeffector Junctions and Peripheral Synapses,* John Wiley and Sons, New York.

Adrenoceptors

Raymond P. Ahlquist

Department of Pharmacology, Medical College of Georgia, Augusta, Georgia 30901, U.S.A.

From figment to fact

The modern adrenoceptors were invented about 30 years ago to explain the different pharmacodynamic effects of some congeners of adrenaline[1]. Why did phenylephrine (Neo-Synephrine) constrict blood vessels? Why did isoprenaline cause vasodilation? Why did adrenaline cause both of these effects? The simplest answer was to imagine that there are two different adrenoceptors. To demonstrate our scientific acumen we named these *alpha* and *beta*.

All adrenergic responses are controlled by one or the other of these receptors. Some of the clinically important responses are as follows:

The *alpha* adrenoceptor is associated with contraction of involuntary (smooth) muscle. Blood vessels are constricted; the radial muscle of the iris is contracted causing mydriasis; the uterus and the spleen are contracted. The smooth muscle of the intestinal tract is relaxed by the *alpha* adrenoceptor.

The *beta* adrenoceptor is associated with relaxation or inhibition of smooth muscle activity. Blood vessels are dilated; bronchial smooth muscle is relaxed; intestinal activity is inhibited. The heart is stimulated by *beta* adrenoceptors. Heart rate is increased, the force of contraction is augmented, and impulse conduction rate is increased.

The adrenoceptors were first delineated in terms of selective agonists Phenylephrine acted mainly through *alpha* adrenoceptors, isoprenaline through *beta* adrenoceptors, and adrenaline and noradrenaline acted through both adrenoceptors.

It took about 10 years for this adrenoceptor concept to be universally accepted. Although this seems a long time, it was a short time when compared with the acceptance of the idea of neurochemical transmission[4]. DuBois Raymond in 1877 first suggested the possibility of neuronal chemical transmission. Adrenergic transmission was almost discovered in 1904. Chemical transmission was established in 1923 by Otto Loewe who found that the vagus nerve when stimulated released acetylcholine. In the 1940s von Euler proved that noradrenaline was the adrenergic neurotransmitter.

Even when chemical transmission was proved, the idea of a receptor was not emphasized. Hormones were said to act on 'target organs'. This implied that hormones sought out the tissue that was to respond. In fact, of course it is the responding tissue that is able to recognize the circulating hormone.

To gain acceptance the dual adrenoceptor idea had to overcome a number of objections. Cannon's concept of two adrenergic transmitters, Sympathin E and I, was the order of the day[5]. The idea of two sympathins was so firmly entrenched that our dual receptor paper was rejected by a well-known pharmacology journal[2]. Adrenergic vasodilation was also unacceptable to many pharmacologists. As late as

1961 it was argued that this vasodilation was a secondary response due to lactic acid produced by a metabolic effect of adrenaline.

The dual adrenoceptor concept resulted from a comparative study of five closely related catecholamines. Similar studies of phenolic compounds did not support the dual receptor idea. It was the discovery of the 'indirect' action of many adrenergic agonists that overcame this objection. These 'indirect' agonists act by entering the nerve terminal and displacing or releasing noradrenaline that then acts on the receptor. However, it was the discovery of beta adrenoceptor blocking agents that finally made the dual adrenoceptor concept palatable.

Within the past few years the physical presence of adrenoceptors has been conclusively demonstrated by radio-ligand binding studies[8].

Drug development

In the early part of this century chemists searched for adrenaline substitutes. Vasoconstriction, bronchodilation, prolonged action and oral effectiveness were looked for. Substitutes for the ergot alkaloids, substances that blocked some of the actions of adrenaline, were also sought. By 1948 most of the drugs now classified as alpha adrenoceptor agonists and blocking agents were known. Isoprenaline, the prototype beta adrenoceptor agonist was introduced in the 1940s by Konzett.

The missing piece in the adrenoceptor puzzle was a drug to block beta adrenoceptors; we searched diligently to find one. The late Bernard Levy almost succeeded. He found that the alpha adrenoceptor agonist, methoxamine, appeared to block some beta responses. Later we found, with the benefit of hindsight, that a compound we had worked with extensively, ethyl–noradrenaline, was a beta blocker.

Dichloroisoprenaline (DCI) the first practical beta adrenoceptor blocking agent was first prepared by Slater and Powell of the Eli Lilly Co. The first definitive study of DCI was done by Moran and Perkins at Emory University in Atlanta, Georgia[6]. Although a difficult compound to work with, being a potent partial agonist, DCI quickly opened up new fields of study in physiology and pharmacology. The demonstration by Black that beta block was effective in angina initiated the search for better beta blockers. The first compound pronetholol appeared to be carcinogenic in mice. Clinical testing of this drug was abandoned, but the stigma of possible carcinogenicity unfortunately continues to be attached to all new beta blockers. Propranolol was then introduced. Because of its long use and thorough study propranolol must be considered the prototype to which all other beta blockers should be compared[3].

At the present time there are about 15 beta blockers available for clinical use throughout the world. The only actions that they all share in common are blockade of the cardiac beta adrenoceptors and blockade of renin release in the kidney. The variable properties can be briefly outlined as follows:

1. Propranolol and most of the others block all beta adrenoceptors. Some, such as metoprolol, act selectively on the cardiac receptors.

2. Propranolol is a potent local anesthetic. Some of the others such as atenolol are not. In the early days of propranolol it got the reputation of being a potent myocardial depressant. This was because it was given in the wrong dose to the wrong patient. Blockade of beta adrenoceptors can be only harmful if sympathetic drive is all that is keeping the heart going. However, considering the vast amount of propranolol given to millions of patients, this beta blocker is obviously not a significant myocardial depressant.

3. Propranolol is lipid soluble. Some of the others are not. Penetration into the

central nervous system is controlled by lipid solubility. Therefore, propranolol can act in the CNS while some of the others cannot.

4. Propranolol is rapidly cleared by the liver. There is a large interpatient variation in this clearance rate. Therefore, it is difficult to predict serum level on the basis of oral dosage. Most of the other *beta* blockers, especially the lipid soluble ones, share this phenomenon. Since this property is not unique to the *beta* blockers it should not be considered any limitation on the therapeutic use. All it means is that there is no single average dose. The dose is the amount that produces the therapeutic action.

5. Propranolol is not a partial agonist. That is, it has no intrinsic agonist action. Some of the others are partial agonists. These drugs may produce cardiac stimulation when first given. Alprenolol and pindolol are examples.

Clinical uses of *beta* blockade

It is not our purpose to describe in detail the clinical applications of *beta* adrenoceptor blocking agents. Since 1963 there have appeared hundreds of papers, proceedings, monographs and books on this subject. It is sufficient to state that these drugs are invaluable in many types of heart disease. However, the antihypertensive action deserves special mention.

The antihypertensive action of the *beta* blocker propranolol was discovered in about 1963 by Prichard[7]. The safety and efficacy of this drug was soon established everywhere except in the United States. Because of the potential multi-million dollar market other *beta* blockers were quickly developed. All *beta* blockers are equally effective in treating essential hypertension. Yet as noted above these drugs have other properties that might make a particular one more useful in some patients. In some countries, notably the U.S., regulatory agencies seem to be trying to control medical practice as shown by the following illustration.

Propranolol entered clinical practice in the United Kingdom in 1965. However, in the U.S. propranolol was approved only for arrhythmias in 1968; for angina pectoris in 1973; and for hypertension in 1976, a delay of 11 years. At the present time nine *beta* blockers are available in Britain, and only three in the U.S. Not only does the U.S. regulatory agency limit clinical use by approving only certain indications; it also controls the size of tablets and the drug information that the drug manufacturer can supply to physicians. Clinical decisions are being impeded by bureaucratic actions.

Recapitulation

The development of the dual adreno-ceptor concept directly or indirectly influenced many aspects of biomedical research.

1. The development of more selective adrenergic agonists and blocking agents was made possible. This, in turn, made available more pharmacologic tools to study control mechanisms in the body.

2. Interest was rekindled in receptors in general. Now we accept receptors for such diverse substances as serotonin, histamine and morphine. This has also stimulated the search for new transmitter chemicals.

3. It is now possible to separate drugs on the basis of whether they act on the receptor or on the biosynthesis, storage or release of the neurotransmitter.

4. The *beta* adrenoceptor was one of the first receptors involved in the development of the second messenger concept. Stimulation of this receptor causes the activation of intracellular adenylate cyclase. This enzyme catalyzes the formation of cyclic-AMP which in turn activates contractile and metabolic mechanisms in the cell. This mechanism is not unique for *beta*

adrenoceptors, since most hormones use the same system.

5. Early studies on the *beta* adrenoceptor revealed that there are subtypes of this receptor. For example the *beta* receptors of the heart are different than those of the bronchi. A practical result of this has been the development of cardio-selective *beta* blockers and broncho-selective *beta* agonists.

6. The effectiveness of *beta* blockers in essential hypertension has opened new areas of study in this disorder. One possible mechanism of action of *beta* blockers in hypertension is the inhibition of renin release in the kidney. The angiotensin resulting from the renin is a good candidate for the cause of the elevated pressure; angiotensin is a vasoconstrictor and also causes sodium retention. Although there is no agreement that *beta* blockers act in hypertension by inhibition of renin, it is interesting to note that another new class of drugs which inhibits angiotensin formation is also antihypertensive.

7. Finally, by using radioactive agonists and blocking agents the adrenoceptors have been isolated and quantitated. Much of the older information about drug-receptor interactions has been verified by these radio-ligand studies. And it is now possible to show the effects of such things as endocrine hormone levels on the number of adrenoceptors.

Reading list

1. Ahlquist, R. P. (1948) *Am. J. Physiol.* 154, 586–99.
2. Ahlquist, R. P. (1973) *Perspec. Biol. Med.* 17, 119–22.
3. Braunwald, E. (ed.) (1978) *Beta-Adrenergic Blockade – A New Era in Cardiovascular Medicine,* Excerpta Medica, Princeton.
4. Breathnach, C. S. (1978) *J. Ir. Coll. Physicians Surg.* 7, 1–8.
5. Cannon, W. B. and Rosenblueth, A. (1937) *Automatic Neuroeffector Systems,* Macmillan, New York.
6. Moran, N. C. and Perkins, M. E. (1958) *J. Pharmacol. Exp. Ther.* 124, 223–37.
7. Prichard, B. N. C. and Gillam, P. M. S. (1964) *Br. Med. J.* 2, 725–28.
8. Williams, L. T. and Lefkowitz, R. J. (1978) *Receptor Binding Studies in Adrenergic Pharmacology,* Raven, New York.

Towards the chemical and functional characterization of the β-adrenergic receptor

A. Donny Strosberg*, Georges Vauquelin†,
Odile Durieu-Trautmann*,
Colette Delavier-Klutchko*, Serge Bottari†
and Claudine Andre†

* *Institut de Recherche en Biologie Moleculaire, Tour 43, 2 Place Jussieu, 75221 Paris, Cedex 05, France.*
† *Institute of Molecular Biology, Free University of Brussels (VUB), Brussels, Belgium.*

The biochemical characterization of the catecholamine β-adrenergic receptor from turkey erythrocyte membranes is rapidly progressing. The complex relationship with the adenylate cyclase and other membrane components which intervene in the hormonal stimulation of the cell is discussed in view of the effect of various ligands on the membrane-bound and affinity-purified receptor.

A large variety of hormones and neuro-transmitters exert their physiological effects by binding to specific receptor sites, located on the external surface of the plasma membranes of the target cells. This binding is followed by the activation of adenylate cyclase on the cytoplasmic side. While the interaction between the hormone and receptor can be easily assessed and defined by measuring the adenylate cyclase activity, direct characterization of receptors as discrete membrane components has been difficult. In fact, for a long time, receptors remained abstract concepts whose existence was proposed only to explain pharmacological effects on target tissues.

A number of models have been proposed to explain the mechanisms of recognition of receptors and activation by hormones[1]. However, the elucidation of the molecular basis of these phenomena will undoubtedly require the solubilization and purification of the various active components of the hormone-responsive adenylate cyclase system, comprising receptor, transmitter and catalytic unit, followed by reconstitution *in vitro* to a hormone-responsive system.

Synthesis of radiolabelled ligands which bind to hormone receptors on the cell surface with high specificity and affinity, along with new developments in methods for solubilizing and purifying membrane proteins have led to a more precise definition of these components. These studies are closely paralleled by advances in the functional and chemical characterization of the adenylate cyclase enzyme and guanine nucleotide-binding components which play a major role in the transfer of the signal between the activated receptor and the enzyme.

This review discusses advances in the purification and chemical characterization of the components intervening in the stimulation of adenylate cyclase by catecholamines.

Molecular characterization of the β-adrenergic receptor

The catecholamine hormones, adrenaline and noradrenaline, and a large number of pharmacologically active analogs have been shown to increase considerably adenylate cyclase activity in cell membranes derived from various sources[2]. This stimulation is stereospecific (the levorotary isomer of the catecholamine hormone being the most active), and occurs via interaction of the ligand with so called β-adrenergic receptors. Catecholamine-stimulated adenylate cyclase activity can be specifically and reversibly inhibited by antagonists, whose sole property is to occupy the β-adrenergic receptors.

Radiolabelled agonists and antagonists allow the direct characterization of β-adrenergic receptors[2]. The model system most used in our laboratory utilizes the turkey erythrocyte membrane. In this system, binding of $(-)-[^3H]$dihydroalprenolol (DHA) occurs to a single class of non-cooperative sites $(0.2-0.3$ pmol/mg membrane protein) with an equilibrium dissociation constant (K_D) of 6 nM[4]. Binding is fast and reversible. At 30°C, equilibrium binding is achieved within 1 min, and dissociation of the bound tracer occurs within the same time upon dilution, or by addition of an excess of the unlabelled antagonist $(-)$-propranolol[4]. Displacement is stereospecific and occurs with the order of potencies: isoproterenol $\simeq$ protokylol $>$ noradrenaline $\simeq$ adrenaline, as measured by adenylate cyclase activation in this system[4].

The characteristics confirm that DHA binding to turkey erythrocyte membranes corresponds to that expected for adenylate cyclase-linked β_1-adrenergic receptor sites.

Presence of one or several disulfide bonds

Both the membrane-bound and purified β-adrenergic receptor from turkey erythrocytes can be inactivated by the reducing agent dithiothreitol[3]. This effect can be inhibited by the preliminary binding of β-adrenergic agonists and antagonists, suggesting that these drugs can effectively protect one or several essential disulfide bonds of the receptor. The protection proceeds either by direct shielding of disulfide bonds located at the binding site (Fig. 1), or by the induction of a conformational change of the receptor resulting in burying of disulfide bonds distant from the binding site.

Conformational change upon binding of β-adrenergic agonists

The effect of the alkylation reagent, N-ethylmaleimide (NEM), on the β-adrenergic receptor differs markedly from the one observed for dithiothreitol. Whereas pretreatment of the membranes with either NEM or the β-adrenergic agonist $(-)$-isoproterenol alone does not affect subsequent DHA binding to receptor sites, the simultaneous presence of both compounds causes a decline of nearly 50% in the number of sites[5]. A striking correlation between the agonist character (i.e. partial activity for adenylate cyclase stimulation) and the rate of alkylation is observed[6]. Antagonists are ineffective in potentiating the alkylation. These results suggest that binding of an agonist induces a change in conformation of the receptor which leads to an increased accessibility of a NEM-sensitive site (Fig. 1). The close correspondence between the ability of β-adrenergic ligands to induce activation of adenylate cyclase and to cause this conformational

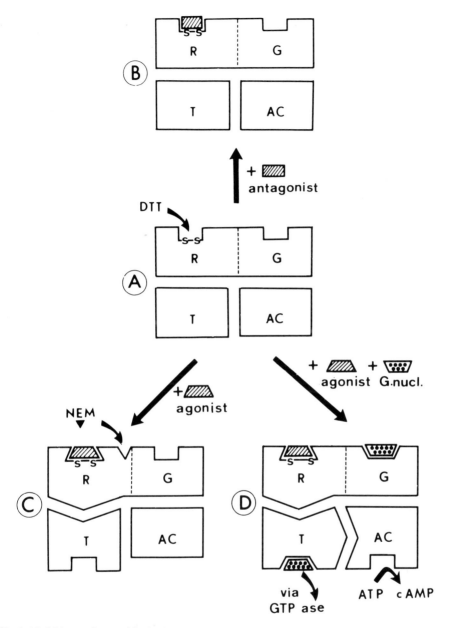

Fig. 1. Model for regulation of the β-adrenergic adenylate cyclase system by β-adrenergic agents and guanine nucleotides. The system is composed of: β-adrenergic receptors (R), transmitter molecules (T) and adenylate cyclase enzymes (AC). Guanine nucleotide regulatory sites (G) are associated, or part of R. The scope of this model is to illustrate the molecular phenomena associated with β-adrenergic stimulation of the system. Stoichiometry, mobility and vectorial location of the different compounds are not considered. (A) No bound ligands: R can be inactivated by the reducing agent dithiothreitol (DTT), indicating the exposure of essential disulfide bonds. (B) β-adrenergic antagonists bound to R: disulfide bonds of R are shielded. (C) β-adrenergic agonists bound to R: disulfide bonds of R are shielded, but R can be inactivated by NEM, indicating the exposure of essential alkylable groups. Interaction between guanine nucleotides and T is favoured. (D) Binding of β-adrenergic agonists to R and guanine nucleotides to T and G: G induces protection of the alkylable groups of R. T induces activation of AC, resulting in the conversion of ATP into cyclic AMP. AC activation is terminated by hydrolysis of the guanine nucleotide bound to T (by the GTPase activity of T).

change further indicate that both phenomena are closely related.

Solubilization and purification of the β-adrenergic receptor

β-Adrenergic receptors from turkey erythrocyte membranes can be solubilized in an active form by treatment with the plant glycoside digitonin[4,7]. Other detergents such as Lubrol PX and WX, Triton X-100 and X-305, sodium deoxycholate, Nonidet P-40, Tween-60 and lithium diiodosalicylate are ineffective. Purification of the solubilized receptors can be achieved by affinity chromatography[4,7]. For this purpose, the digitonin extract of turkey erythrocyte membranes is loaded on an affinity column consisting of alprenolol linked via a hydrophilic spacer arm to agarose beads. The receptors are retained while other solubilized membrane proteins, including the adenylate cyclase enzyme, pass through. After washing, application of an excess of free ligand causes release of the purified receptor molecules from the gel. Using this affinity chromatography technique, we have achieved a 12,000-fold purification of the turkey erythrocyte β-adrenergic receptors with retention of all the original (−)-[³H]dihydroalprenolol binding properties. A recent report of the purification of the β-adrenergic receptor from frog erythrocyte membranes[8] essentially confirmed our results.

Molecular weight of the receptor and its subunits

The unpurified, digitonin-solubilized β-adrenergic receptor from turkey erythrocyte membranes has a mol. wt in the range of 200,000 when assessed by gel filtration (unpublished observations). The mol. wt of the receptor from frog erythrocytes was estimated as 150,000[9] and that of the Lubrol-solubilized receptor from S49 lymphoma cells was estimated by sucrose density gradient centrifugation to be

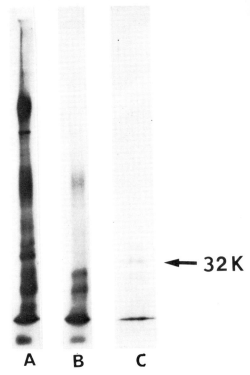

Fig. 2. SDS-polyacrylamide gel electrophoresis of the β-adrenergic receptor from turkey erythrocyte membranes. Autoradiography of: iodinated proteins from digitonin-solubilized membranes applied to (A), directly eluted from (B), and biospecifically eluted (C) from the affinity gel presented in Fig. 3. The major receptor component seen in (C) does not appear when the membrane proteins are applied on the affinity gel in the presence of the antagonist (−)-propranolol or of specific rabbit antireceptor antibodies.

70,000[10]. By SDS-polyacrylamide gel electrophoresis and affinity labelling of the β-adrenergic receptor, a tentative identification of two subunits with mol. wts of 37,000 and 41,000 was reported for the receptors obtained from rat skeletal myoblasts grown in culture and from turkey erythrocytes[11]. After iodination of turkey erythrocyte membranes, the β-adrenergic receptors (purified by affinity chromatography) were analysed by SDS-polyacrylamide gel electrophoresis followed by autoradiography; we found a major component of mol. wt 32,000 (submitted for publication; Fig. 2). This component is not revealed when the affinity chromatography is performed in the presence of an excess of free antagonist, propranolol, or in the presence of specific anti-receptor antibodies. The mol. wt of the major component remains the same when the samples are treated with β-mercaptoethanol prior to SDS-polyacryamide gel electrophoresis.

Comparison of the mol. wts of the β-adrenergic receptors obtained by gel filtration or sucrose density centrifugation with values obtained by SDS-polyacrylamide gel electrophoresis, strongly suggests that the receptors are composed of multiple subunits.

Components of the catecholamine-sensitive adenylate cyclase system

The catecholamine-sensitive adenylate cyclase system is composed of at least three distinct components: the β-adrenergic receptor, the adenylate cyclase enzyme and a transmitter molecule that binds guanine nucleotide[12]. The existence of this latter component is deduced from the observation that catecholamine hormones, as well as several other hormones, stimulate the system only in the presence of guanine nucleotides. Transient activation in the presence of GTP, compared to persistent activation in the presence of the non-hydrolysable analog guanyl-5'-yl imido-diphosphate (Gpp(NH)p) further suggests that this transmitter molecule possesses a GTPase activity[12].

It is now generally believed that the β-adrenergic receptor does not cause direct adenylate cyclase activation. Instead, the catecholamine-receptor complex probably favors the interaction between the guanine nucleotide and the transmitter molecule; the neocleotide–transmitter complex being responsible for the activation of adenylate cyclase[12] (Fig. 1). The recent finding of a catecholamine-stimulated GTPase activity in turkey erythrocyte membranes by Cassel and Selinger[13], which has been confirmed

in our laboratory (submitted for publication) strongly argues for such a model.

Resolution and isolation of the different components

Orly and Schramm demonstrated the functional individuality of the β-adrenergic receptor and the adenylate cyclase[14]. By fusing Friend erythroleukemia cells lacking receptor sites and chemically treated turkey erythrocytes with no residual adenylate cyclase activity, these authors showed that a catecholamine-sensitive adenylate cyclase activity could be restored. The receptor and enzyme thus behave as distinct components which are capable of individual migration in the membrane. Resolution of the β-adrenergic receptor from the adenylate cyclase enzyme has also been achieved. Both the receptor and the enzyme from turkey and frog erythrocyte plasma membranes can be solubilized by treatment with the detergent, digitonin. These components can be partially separated by gel filtration[15] and completely separated by affinity chromatography[4] (Fig. 3). The adenylate cyclase enzyme from dog myocardial membranes was considerably purified using a combination of hydrophobic chromatography and affinity chromatography on ATP-Sepharose[16]. The 5000-fold purified protein was still sensitive to Gpp(NH)p stimulation, suggesting that the transmitter component remains attached to the enzyme during this procedure.

Individuality of the transmitter molecules and the enzymes has been demonstrated for a number of systems. The GTP-binding proteins from solubilized pigeon erythrocyte membranes were separated from adenylate cyclase by affinity chromatography on a GTP-Sepharose matrix[17]. The specifically eluted regulatory protein restored Gpp(NH)p and NaF-stimulated adenylate cyclase activity to a preparation deprived of the guanine nucleotide-binding proteins. Treatment of

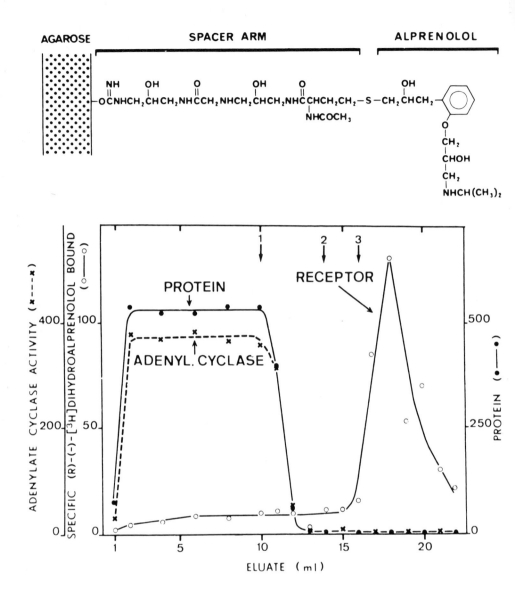

Fig. 3. Affinity chromatography purification of the β-adrenergic receptor from turkey erythrocyte membranes. The structure of the thio-alprenolol side arm grafted on the Sepharose 4B gel is presented in the upper part of the figure. The digitonin-solubilized membranes, applied to the column, contain active adenylate cyclase which is not retained. 95% of the solubilized receptors are retained, and after washing of the column (arrows 1 and 2), eluted in presence of free (−)-[³H]dihydroalprenolol (arrow 3).

solubilized S49 lymphoma cell membranes with NEM causes selective inactivation of the adenylate cyclase enzyme[18]. The remaining soluble transmitter component is still functional when re-incorporated in membranes of a phenotypically trans-mitter-deficient S49 lymphoma cell variant which still contains β-adrenergic receptors and adenylate cyclase.

Guanine nucleotide regulation of β-adrenergic receptors

A growing list of observations indicates that guanine nucleotides also affect the structural and functional properties of the β-adrenergic receptor. These compounds markedly reduce the affinity of β-adrenergic agonists for their receptor in frog erythrocyte[19] and S49 lymphoma cell membranes[20]. Agonists also desensitize approximately 50% of the receptor sites in the frog erythrocyte system by converting them to a 'high affinity' state[21]. Subsequent addition of guanine nucleotides reverses this desensitization process[19,21]. Neither of these phenomena is observed for the β-adrenergic receptor in turkey erythrocyte membranes. However, in this system, both GTP and Gpp(NH)p confer an effective protection of the agonist-bound receptor against inactivation by NEM (submitted for publication). These three phenomena have an interesting point in common: i.e. the effect of the non-hydrolysable Gpp(NH)p is reversible at the level of the receptor compared to the persistent adenylate cyclase stimulation by the same nucleotide. Although final evidence will have to await the complete resolution of the catecholamine-sensitive adenylate cyclase system, these data suggest that β-adrenergic receptors are under the control of a guanine nucleotide-binding component which is functionally distinct from the transmitter molecule (Fig. 1).

Conclusion

The β-adrenergic adenylate cyclase system is composed of at least three main components: the receptor, the transmitter and the adenylate cyclase catalytic unit. The existence of a fourth component by which guanine nucleotides affect the receptor is still hypothetical (Fig. 1).

The receptor binds catecholamine agonists and antagonists. Binding either of these ligands effectively protects against reduction by dithiothreitol – one or several disulfide bonds being essential for the receptor-binding capability. The interaction of the receptor with agonists results in a change of conformation which renders it sensitive to the alkylating agent NEM. This same change in conformation or an 'anchoring effect' caused by the agonist binding may constitute the signal which results, through the transmitter, in activation of the adenylate cyclase catalytic unit. The transmitter is a regulatory protein whose activity is modulated by the nucleotide GTP, its synthetic analog Gpp(NH)p, or NaF. The cyclase catalytic unit is inactivated irreversibly by NEM.

Each of these three components of the β-adrenergic system can now be solubilized and isolated in a functionally active form, free of the biological activity corresponding to the other interacting proteins of the system. The chemical characterization of the receptor has been undertaken despite the very small amounts of protein available. Recent progress in the development of micro-sequence methodology has allowed us to determine the amino terminal sequence of the acetylcholine receptor, a 40,000 mol. wt subunit from the electric *Torpedo electroplax*[22]. In the near future, we hope to undertake structural studies of the β-adrenergic receptor as well.

Acknowledgements

We thank Dr J. D. Capra (Dallas) and Dr J. Hoebeke (Brussels) for critical review and Dr J. Hanoune (Créteil) and Dr P. Janssen (Beerse) for helpful discussions. Georges Vauquelin is Aangesteld Navorser of the National Fonds voor Wetenshappelijk Onderzoek, Belgium. The work described here was supported by grants from the NFWO, FGWO, IWONL and Janssen Pharmaceutica (Belgium) and by CNRS and INSERM (France).

Reading list

1 Greaves, M. F. (1977) *Nature (London)* 265, 681–683
2 Haber, E. and Wrenn, S. (1976) *Physiol. Rev.* 56, 317–338
3 Vauquelin, G., Bottari, S., Kanarak, L. and Strosberg, A. D. (1979) *J. Biol. Chem.* 254, 4462–4469

4 Vauquelin, G., Geynet, P., Hanoune, J. and
   Strosberg, A. D. (1977) *Proc. Natl. Acad. Sci.
   U.S.A.* 74, 3710–3714
5 Bottari, S., Vauquelin, G., Durieu, I., Klutchko,
   C. and Strosberg, A. D. (1979) *Biochem. Biophys.
   Res. Commun.* 86, 1311–1318
6 Vauquelin, G., Bottari, S. and Strosberg, A. D.
   (1979) *Mol Pharmacol.* (in press)
7 Vauquelin, G., Geynet, P., Hanoune, J. and
   Strosberg, A. D. (1979) *Eur. J. Biochem.* 98,
   543–556
8 Caron, M. G., Srinivasan, Y., Pitha, J., Kociolek,
   K. and Lefkowitz, R. J. (1979) *J. Biol. Chem.* 254,
   2923–2927
9 Caron, M. G. and Lefkowitz, R. J. (1976) *J. Biol.
   Chem.* 251, 2374–2384
10 Haga, T., Haga, K. and Gilman, A. G. (1977) *J.
   Biol. Chem.* 252, 5776–5782
11 Atlas, D. and Levitzki, A. (1978) *Nature (London)* 272, 370–371
12 Levitzki, A. and Helmreich, E. J. M. (1979) *FEBS
   Lett.* 101, 213–219
13 Cassel, D. and Selinger, Z. (1976) *Biochim. Biophys. Acta* 452, 538–551
14 Orly, J. and Schramm (1976) *Proc. Natl. Acad.
   Sci. U.S.A.* 73, 4410–4414
15 Limbird, L. E. and Lefkowitz, R. J. (1977) *J. Biol.
   Chem.* 252, 799–802
16 Homcy, C., Wrenn, S. and Haber, E. (1978) *Proc.
   Natl. Acad. Sci. U.S.A.* 75, 59–63
17 Pfeuffer, T. (1977) *J. Biol. Chem.* 252,
   7224–7234
18 Ross, E. M., Howlett, A. C., Ferguson, M. and
   Gilman, A. G. (1978) *J. Biol. Chem.* 253,
   6401–6412
19 Mukherjee, C. and Lefkowitz, R. J. (1977) *Mol.
   Pharmacol.* 13, 291–303
20 Maguire, M. E., Van Arsdale, P. and Gilman,
   A. G. (1976) *Mol. Pharmacol.* 12, 335–339
21 Williams, L. T. and Lefkowitz, R. J. (1977) *J. Biol.
   Chem.* 252, 7207–7213
22 Devillers-Thiéry, A., Changeux, J. P., Paroutaud,
   P. and Strosberg, A. D. (1979) *FEBS Lett.* 104,
   99–105

New directions in adrenergic receptor research

Part I

Robert J. Lefkowitz and Brian B. Hoffman

The Howard Hughes Medical Institute Laboratory, Departments of Medicine (Cardiovascular) and Biochemistry, Duke University Medical Center, Durham, NC 27710, U.S.A.

The direct identification of drug receptors by radioligand binding studies has led to a virtual revolution in research in the area of adrenergic pharmacology over the past five years. Information about the alpha- and beta-adrenergic receptors, made possible by such ligand binding studies, has been accumulating at an ever quickening pace and a number of comprehensive reviews in this area have appeared over the past few years[1-3]. Consequently, it is not our purpose in this series of articles to review the area of ligand binding to adrenergic receptors exhaustively. Rather, we attempt to focus on trends and developments of the past year or two which seems to us particularly exciting and potentially likely to lead to new insights about the adrenergic receptors. The first part of this article emphasizes the beta-adrenergic receptors and the second part, the alpha-adrenergic receptors.

Ahlquist initially classified the adrenergic receptors into two major categories: alpha and beta. This classification was based on the characteristic relative potency orders of a variety of adrenergic agonists. Typically, for β-adrenergic responses such as positive inotropism in the heart, isoproterenol is a more potent amine than epinephrine which in turn is equipotent with or more potent than norepinephrine. Conversely, for α-adrenergic-mediated responses such as smooth muscle contraction, epinephrine is generally the most potent followed by norepinephrine with isoproterenol relatively very weakly potent. There are highly specific antagonists for both beta receptors and alpha receptors, exemplified by propranolol and phentolamine respectively. Both receptors are characterized by marked stereospecificity with the (l) or ($-$) stereoisomers of both agonists and antagonists being considerably more potent than the (d) or ($+$) stereoisomers.

There appear to be at least two major subtypes of each of the adrenergic receptors. β_1- and β_2-adrenergic receptors differ in their relative potencies for epinephrine and norepinephrine[4]. At β_1-adrenergic receptors, such as those in cardiac and adipose tissue, epinephrine and norepinephrine appear to be about equally potent. By contrast, at β_2-adrenergic receptors, such as those in smooth muscle, epinephrine is, in general, at least 10 times more potent than norepinephrine. Both the β_1- and β_2-adrenergic receptors are in general coupled to the membrane-bound enzyme adenylate cyclase and their occupancy by agonists leads to stimulation of this enzyme with consequent elevation of the intracellular concentrations of the second messenger cyclic AMP.

There are also subtypes of α-adrenergic receptors. α_1-Adrenergic receptors are the typical postsynaptic α receptors which mediate, for example, the classical smooth muscle constricting effects of α-adrenergic agonists. In addition, there appears to be another class of α-adrenergic receptors. These were initially discerned based on physiological experiments which indicated that catecholamines served, in an autoinhibitory fashion, to inhibit their own release from sympathetic nerve terminals. This autoinhibitory effect, mediated at the level of the presynaptic nerve terminals, appears to be mediated by α-adrenergic receptors. However, certain compounds are considerably more potent at these so-called presynaptic receptors than at the classical postsynaptic α-receptors. Examples of such α_2 selective drugs are yohimbine, an antagonist, and clonidine, an agonist. It is now clear that although the presynaptic α receptors share many properties with α_1 or postsynaptic adrenergic receptors there are differences in their specificities. Accordingly, these receptors are generally termed 'alpha$_2$' or 'presynaptic'. Since α receptors having characteristics pharmacologically very similar to those of the so-called presynaptic receptors are also found postsynaptically, it has become popular, over the past several years to refer to such receptors as 'alpha$_2$' rather than as 'presynaptic'. An example of an α_2 receptor which is not presynaptic is the α_2-adrenergic receptor which inhibits adenylate cyclase in human platelets. In general, it seems likely that α_1-adrenergic receptors are not coupled to adenylate cyclase. By contrast, the α_2-adrenergic receptors often appear to inhibit adenylate cyclase activity directly. To date, the most useful drugs for discriminating α_1- and α_2-adrenergic receptors have been the antagonists prazosin and yohimbine. Prazosin is a much more potent antagonist of α_1 than α_2 responses, whereas yohimbine is somewhat more potent at α_2 than α_1 receptors.

A variety of radiolabelled adrenergic agents have been used successfully for binding studies of the β-adrenergic receptors. These are listed in Table I. All four of the available antagonist agents are not subtype selective. That is, they appear to bind with approximately equal affinity to β_1- and β_2-adrenergic receptors. However, the agonists [³H]hydroxybenzylisoproterenol and [³H]epinephrine appear to label the β_2-adrenergic receptors somewhat selectively. There have also been a few reports of the successful labelling of the β receptors with the agonist [³H]isoproterenol, but thus far this does not appear to have been as productive an approach.

Historically, it is worth noting that

TABLE I. Radioligands used for beta-adrenergic receptor identification.

Radioligand	Receptor specificity	Comments
(−)-[³H]Dihydroalprenolol (±)-[¹²⁵I]Hydroxybenzylpindolol (±)-[³H]Propranolol (±)-[³H]Carazolol	Equal affinity at β_1 and β_2 receptors	Antagonists; label entire β-receptor population
(±)-[³H]Hydroxybenzylisoproterenol (−)-[³H]Epinephrine	β_2 High affinity state	Agonists; probably label only the guanine nucleotide sensitive form of the β_2 receptor

early in the 1970s, initial attempts to label β-adrenergic receptors with [³H]-labelled catecholamines led to the labelling of sites which did not possess all the characteristics of true β-adrenergic receptors. However, all of the agents noted in Table I have been shown to label sites which, under appropriate experimental conditions, obey the strict criteria of specificity and stereospecificity expected of β-adrenergic receptors. It is also worth stressing that virtually all of these drugs can lead to labelling of sites other than β receptors if experimental conditions are not correctly chosen or if any of a variety of other technical artifacts supervene.

Beta-adrenergic receptor subtypes

The delineation and quantification of receptor subtypes is a particularly good illustration of the power of ligand binding techniques. This area has been attracting increasing interest in the β-adrenergic receptor field over the past year. Nahorski and colleagues were the first to call attention to the important fact that the competition curves of β-adrenergic subtype selective antagonists in competition with a non-subtype selective radioligand, such as [³H]dihydroalprenolol ([³H]DHA), are complex and sometimes biphasic as opposed to the steep and uniphasic curves obtained with non-subtype selective unlabelled agents[5]. The complex shape of binding curves of the subtype selective agents is due to the fact that they have differential affinity for the two populations of receptors which are being labelled by the radioligand. Thus, there will be two discrete components to their competition curves. Nahorski and colleagues developed a graphical method for estimating the proportions of the receptor subtypes and their respective affinities for the β_1- and β_2-adrenergic receptors which involved a 'pseudo Scatchard' transformation of the data, in which 'percent inhibition of binding' was equated with

the 'bound' term of typical Scatchard plot ('bound/free' versus 'bound'). However, this graphical approach may give erroneous parameter estimates because of several shortcomings. These include: (1) the implicit assumption that the percent displacement of radioactive ligand is equivalent to receptor occupancy, which is true under some but not all experimental conditions; (2) inaccuracies generated by the data transformation procedures themselves and the oversimplification that the two components of such a complex curve can be simply resolved by graphical extrapolation; and (3) the failure to validate by an objective statistical test whether a one-site, two-site or more complex model is actually most appropriate. For example, the utilization of asymptotes to the lines as used in this method relies too heavily on the least accurate points of the curve and underestimates the influence of one class of sites on ligand binding to the other class of sites. In general, curve fitting methods of the 'pseudo Scatchard' type become inaccurate because of the nature of the data transformation techniques in which numbers subject to experimental error are divided by the low concentrations of competitor to produce the graphical coordinates. These transformed values may greatly magnify even small errors in the original data and can produce large fluctuations in the shape of the fitted curve.

Minneman and Molinoff reported a computer-assisted method of obtaining refined parameter estimates which, none the less, relies on essentially the same graphical approach[6]. However, rather than deriving the parameters by graphical inspection, these workers used an iterative computer program to refine the values for the individual parameters. However, many of the same objections which can be raised about the graphical techniques described above apply to this method as well.

Perhaps the best way to approach the quantitative analysis of receptor subtypes is by the use of non-linear least squares curve fitting techniques such as those developed by André DeLean[7] and David Rodbard. Such methods have been now applied to the β-adrenergic receptor as well as the α-adrenergic receptor. There are several advantages of such methods. Fewer assumptions are required, the most important of which is that the law of mass action applies to the binding of multiple ligands to multiple classes of binding sites. The data may be analysed in the original form (amount of radioligand bound versus total concentration of competitor); no transformation of the data is required. Also, no assumptions are made concerning the equivalency of 'percentage inhibition of binding' and receptor occupancy. Rather, receptor occupancy is actually determined based on the concentration of receptors and the concentrations of the various ligands present in the assay system. It is worth pointing out that the method referred to here is basically a computer modelling technique in which the best fit curve for a set of experimental data points is generated based on the law of mass action. This modelling approach also includes the capacity simultaneously to fit several curves together. Thus, the data for several ligands of differing selectivity can be fitted together to determine more accurately the proportions of receptor subtypes present in a membrane preparation. Another advantage of these methods is that statistical tests are built into the programs to evaluate whether a one-site, two-site or more complex 'fit' is most appropriate. For example, a two-site fit for a given set of data points is not considered appropriate unless it leads to a statistically significant improvement in the fit of the data points. This is an important feature. For example, the other two methods described above start with the *assumption* that there are exactly two sites present and then determine their proportions. With use of the curve fitting methods developed by DeLean and Rodbard, there is at least an objective statistical method for evaluating whether the assumption that two receptor subtypes are present is more valid than simply assuming that a homogeneous population of sites is present. These curve fitting programs are available for general distribution.

Using methods such as those described above several groups have confirmed the β-adrenergic subtype selectivity of a variety of antagonists. In general the selectivity for compounds such as practolol, metoprolol or butoxamine is relatively modest and in no case exceeds ten to one hundred fold greater affinity for one or the other receptor subtype. There are other drugs which have much greater selectivity for the α receptor subtypes. α-Receptor subtypes may be studied by analogous methods as will be illustrated in part II of our discussion.

Guanine nucleotide regulation of adenylate cyclase coupled beta-adrenergic receptors

Since the original observations by Rodbell and colleagues some ten years ago[8], it has become clear that virtually all hormone-sensitive adenylate cyclase systems are characterized by the fact that guanine nucleotides, such as GTP or one of its nonhydrolysable analogs such as GMP-P(NH)P, are required for hormone stimulation of the enzyme. It has also been known since 1976 that guanine nucleotides cause an agonist specific reduction in the affinity of β-adrenergic agents for receptors as assessed by ligand binding studies[9,10]. Thus, nucleotides reduce the affinities of agonists but not antagonists for the β-adrenergic receptors. Similar observations have been reported over the past few years for a variety of other neurotransmitter receptors. Insight into the significance of these observations has recently come from application of the same computer modelling techniques discussed above in the context of receptor subtypes. An example is provided by the data shown in Fig. 1, A and B[11]. These are computer-fitted curves for the competition of an antagonist alprenolol and an agonist isoproterenol with the ligand [³H]DHA (an antagonist) for binding to β-adrenergic receptors in frog erythrocyte membranes. The curve for the antagonist alprenolol, as well as that for all other antagonists tested, models best to a single class of site having a single homogeneous affinity for the receptors. By contrast, the curve for the full agonist isoproterenol models significantly better to two states of interaction, having high (K_H) and low (K_L) affinity for the receptors. Approximately three-quarters of the receptors are found in the high affinity form when isoproterenol is present. The ratio of the dissociation constants K_L/K_H is about 60/1 in the experiment shown. The pattern shown for isoproterenol is typical of full agonists. In the presence of high concentrations of guanine nucleotide, the competition curve is shifted to the right and becomes steeper. There is no change in the total number of receptors present. However, now all the receptors display homogeneous low affinity for the agonist. This low affinity appears to be indistinguishable from the lower of the two affinities for isoproterenol observed in the absence of guanine nucleotide. Qualitatively similar patterns are found for partial agonist drugs. However, for the partial agonists which do not cause full stimulation of the enzyme adenylate cyclase, the ratio K_L/K_H is smaller and the percentage of receptors found in the high affinity form is also less. When a spectrum of drugs of graded intrinsic activity is examined it has been found that there is an excellent correlation between the percentage of the receptors in the high affinity form, or the ratio of the two affinity constants K_L/K_H and the intrinsic activity of the drug. Thus, the unique property of agonists as opposed to antagonists, is the ability of the agonists to induce or stabilize a high affinity guanine nucleotide sensitive state of the β-adrenergic receptors in proportion to their intrinsic activity. Other experiments have suggested that this high affinity state is an essential intermediate in the activation of adenylate cyclase.

Work from a number of laboratories over the past year has established that the various regulatory effects of guanine nucleotides on adenylate cyclase and on receptor binding appear to be mediated by a distinct regulatory component of the system. This is generally termed the guanine nucleotide regulatory protein (N-protein), as originally proposed by Rodbell and his colleagues at the NIH. Together with the hormone receptors and the catalytic moiety of the enzyme this protein determines the functioning of the adenylate cyclase system. Although this field is evolving rapidly the current understanding of many investigators may be summarized as follows (Fig. 2): interaction of a hormone agonist with the receptor leads to the initial formation of a lower affinity, freely reversible complex of the hormone with the receptor (HR). When the drug is an agonist some conformational change is induced in the receptor such that the receptor becomes coupled to the guanine nucleotide regulatory protein. A ternary complex of hormone, receptor and N-protein is thus established (HRN). This is presumably the 'high affinity state' of the receptor discussed above. Formation of this complex is associated with a loss of tightly bound GDP from the nucleotide regulatory protein and a facilitation of the binding of the nucleotide GTP to the

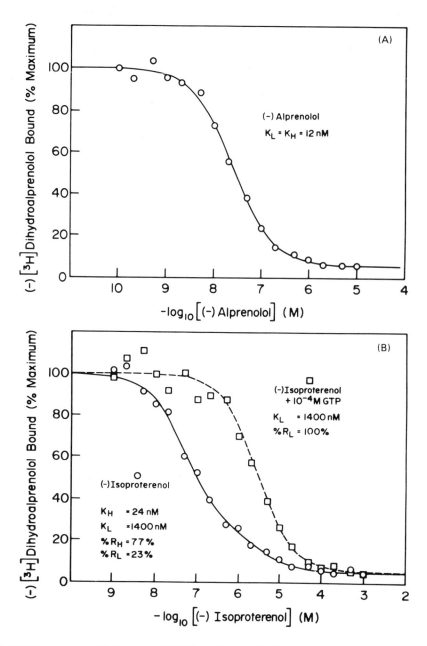

Fig. 1. (A) Computer-modelled curve for inhibition of (−)-[³H]DHA binding to frog erythrocyte membrane
β-adrenergic receptors by (−)-alprenolol. The data points are from a typical experiment. The curve is drawn
by computer modelling procedures for a model with a single binding affinity state for (−)-alprenolol of
12nM. Attempts to fit these binding data to a two-state model did not improve the fit. K_L = low affinity
dissociation constant. K_H = high affinity dissociation constant. Data taken from Ref. 11. (B) Computer-
modelled curves for inhibition of (−)-[³H]DHA binding to frog erythrocyte membrane β-adrenergic
receptors by (−)-isoproterenol in the presence and absence of GTP. The curve for isoproterenol alone was a
significantly better fit ($p < .01$) for a two-state binding model, whereas in the presence of GTP a one-state
model was adequate to fit the data points. K_H and K_L are as defined in (A). % R_H = % receptors in high
affinity state, % R_L = % receptors in low affinity state. The two curves were found to have values for K_L
that were not significantly different. Data taken from Ref. 11.

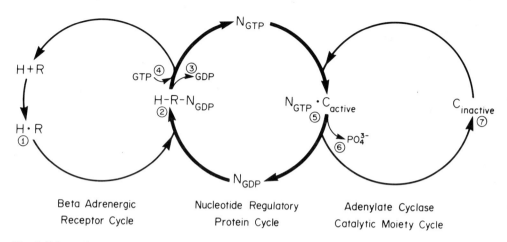

Fig. 2. Schema for a proposed mechanism of adenylate cyclase activation by beta-adrenergic agonists, and GTP. H = hormone or agonist, R = β-adrenergic receptor, N = nucleotide regulatory protein, C = catalytic moiety of adenylate cyclase.

nucleotide regulatory site. Interaction of GTP with this site destabilizes the complex, freeing up the hormone receptor and releasing free nucleotide regulatory protein now charged with GTP. The complex of GTP and nucleotide regulatory protein presumably interacts with the dormant catalytic moiety of the system thereby activating it. Recently Pfeuffer has presented evidence for such a physical association of N-protein and adenylate cyclase [12]. Limbird *et al.*[13] have presented evidence for the physical coupling of the receptor with the N-protein. Once the cyclase is activated, the GTP is cleaved back to GDP by a GTPase activity which resides on the guanine nucleotide regulatory protein[14]. The system is then returned to its baseline state and the cycles may begin again.

As depicted in Fig. 2, what basically seems to be going on are several cycles of interlocking activation and deactivation; one involving hormone receptor and N-protein and the other involving N-protein and adenylate cyclase. These two cycles are connected by the N-protein which acts as a true 'coupling' protein or shuttle between the receptor cycle and the adenylate cyclase cycle. Thus, as originally suggested by Rodbell and his colleagues, the nucleotide regulatory protein may well be the coupler of the receptor and the catalytic moiety.

Recently DeLean *et al.*[15] have demonstrated that the binding of β-adrenergic agonists and antagonists to receptors in frog erythrocyte membranes can be closely simulated and, in fact, fit well to a model involving the stabilization of a *ternary complex* of hormone receptor and an extra membrane component by agonists but not antagonists. The 'extra membrane component' is hypothesized to be the N-protein.

Physiological regulation of beta-adrenergic receptors

One of the most important insights to come from ligand binding studies of adrenergic receptors over the past few years is that the receptors are not static entities in the plasma membranes but rather are subject to very dynamic regulation by a wide variety of physiological and pathophysiological interventions. Table II lists some of the factors which have been found to regulate the β-adrenergic receptors. Among the most significant are the catecholamines themselves which tend to decrease receptor number and biological responsiveness to their own actions. Conversely, both denervation, which leads to a reduction in catecholamine concentration at the β-adrenergic receptor, as well as antagonist treatment, in some circumstances, appear to lead to an increase in receptor number. This increase is possibly accompanied by supersensitivity to agonist action. As noted in Table II, a wide variety of other influences including thyroid hormones and a variety of steroid hormone treatments lead to significant changes in the β-adrenergic receptor content of tissues.

Of the various receptor regulatory influences, perhaps the most intensively studied over the past year or two has been the phenomenon termed 'desensitization'. This term refers to the fact that treatment of tissues or cultured cell systems for a period of time (ranging from minutes to hours) with β-adrenergic agonists leads to a reduction in the tissue's subsequent responsiveness to further β-adrenergic stimulation. When exclusively β-adrenergic stimulation is reduced, with responses to other hormones left intact, the phenomenon is called 'homologous desensitization'; whereas, when responsiveness to all hormonal activators is similarly blunted, the term 'heterologous desensitization' is used[16]. There is increasing evidence that homologous desensitization may be due to a variety of alterations induced in the receptors themselves by β-adrenergic agonist stimulation. By contrast, heterologous desensitization may be caused by alterations at post-receptor steps.

There are at least two types of alterations in β-adrenergic receptors which can be induced by agonists. The first is an 'uncoupling' of the receptors from interactions with more distal components such as, for example, the nucleotide regulatory site. 'Uncoupled receptors' can still be assayed by antagonist binding techniques but will not bind radiolabelled agonists with high affinity. Such 'uncoupled' receptors can also be shown by computer modelling of competition curves (by the methods discussed above) to display lower overall affinity for the agonists[11]. The second defect is an actual loss of receptor binding capacity as assayed by radiolabelled antagonists. The fate of such 'missing' receptors is not clear at present. In analogy with the results reported with other hormone receptors, the β-adrenergic receptors might possibly be internalized. However, it is also possible that they are covalently modified or perturbed in some other way while still remaining in the plasma membrane. In some systems the 'uncoupling' of receptors appears to precede the actual loss of receptor binding capacity but this is not universally the case. The mechanistic relationship between these two types of receptor alterations, i.e. 'uncoupling' and 'receptor loss', remains to be determined.

TABLE II. Some factors regulating beta-adrenergic receptor binding sites.

Regulatory agent or condition	Effect on receptor number
β-Adrenergic catecholamines	Decrease
Propranolol	Increase
Denervation (6-hydroxydopamine, guanethidine, surgical)	Increase
Hyperthyroidism	Increase
Hypothyroidism	Decrease
Cortisone	Decrease (rat liver)
Malignant transformation	Increase
Spontaneously hypertensive rat	Decrease
Alcohol withdrawal	Increase

Solubilization and purification of beta-adrenergic receptors

One of the long range goals in receptor research is, of course, to purify the receptors and learn the intimate molecular details of their structure. For the adenylate cyclase-coupled receptors, such as the β-adrenergic receptor, the task of receptor purification is indeed a challenging one. This is because the receptors are present in minute quantities in plasma membranes and as they are membrane bound they need to be solubilized before purification can begin. To date, unfortunately, no source of β-adrenergic receptors has been found which is as rich as, for example, the electric eels for the cholinergic receptors. None the less, slow but significant progress is being made in β receptor purification. Thus far, most success has been achieved with membranes prepared from avian or amphibian erythrocytes[17,18]. The receptors seem to be efficiently solubilized with the plant glycoside digitonin. After solubilization, the receptors may be purified by affinity chromatography. The best results thus far have been obtained using affinity supports consisting of the β-adrenergic antagonist alprenolol covalently linked to Sepharose[17]. Using such affinity matrices purification of a thousand-fold or more in a single step can be achieved. By combining several steps, it seems likely that it will be possible to purify reasonable quantities of the β-adrenergic receptors in the near future. After this has been accomplished, the detailed biochemical characteristics of the receptor protein can be learned. At present all that is known with confidence is that the receptors are integral membrane proteins distinct from the catalytic moiety of adenylate cyclase. The details of subunit structure etc. remain to be determined.

Clinical studies

A potentially exciting direction for future research in the adrenergic receptor area is the application of ligand binding techniques to the unravelling of pathophysiological alterations in the receptors which might be of clinical importance. In analogy with insulin receptors, it was found several years ago that β-adrenergic receptors were present on human lymphocytes and polymorphonuclear leukocytes. These receptors are coupled to adenylate cyclase and appear to have characteristics quite similar to those of β₂-adrenergic receptors prepared from a variety of other mammalian and non-mammalian sources. Since such circulating cells can be readily obtained by venipuncture, many investigators have become interested in the possibility of using them to learn the status of β-adrenergic receptors in patients. It has already been demonstrated, for example, that the β receptors in such cells display a desensitization analogous to that found in other animal cells. Investigators have examined whether these leukocyte β-adrenergic receptors might be altered in such conditions as asthma, atopic diseases, and thyroid disease, among others. The potential importance or validity of this approach is as yet uncertain. It is worth underscoring, however, that before one can accept the results of such studies as being representative of the status of β-adrenergic receptors in other tissues, it will need to be carefully documented that the receptors in leukocytes do, in fact, show regulatory properties similar to those of the receptors in tissues such as the heart and lungs. This approach, however, remains a potentially promising one for investigating adrenergic receptor regulation *in vivo* in man.

Reading list

1 Wolfe, B. B., Harden, T. K. and Molinoff, P. B. (1977) *Annu. Rev. Pharmacol. Toxicol.* 17, 575–604

2 Maguire , M. E., Ross, E. M. and Gilman, A. G. (1977) *Adv. Cyclic Nucleotide Res.* 8, 1–83

3 Hoffman, B. B. and Lefkowitz, R. J. (1980) *Annu. Rev. Pharmacol. Toxicol.* 20, 581–608

4 Lands, A. M., Arnold, A., McAuliff, J. P., Luduena, F. P. and Brown, T. C. (1967) *Nature (London),* 214, 597–598

5 Barnett, O. B., Rugg, E. L. and Nahorski, S. R. (1978) *Nature (London)* 273, 166–168

6 Minneman, K. P., Dibner, M. D., Wolfe, P. B. and Molinoff, P. B. (1979) *Science* 204, 866–868

7 Hancock, A. A., DeLean, A. L. and Lefkowitz, R. J. (1979) *Mol. Pharmacol.* 16, 1–9

8 Rodbell, M., Birnbaumer, L., Pohl, S. L. and Krans, H. M. J. (1971) *J. Biol. Chem.* 246, 1877–1882

9 Maguire, M. E., VanArsdale, P. M. and Gilman, A. G. (1976) *Mol. Pharmacol.* 12, 335–339

10 Lefkowitz, R. J., Mullikin, D. and Caron, M. G. (1976) *J. Biol. Chem.* 251, 4686–4692

11 Kent, R. S., DeLean, A. and Lefkowitz, R. J. (1980) *Mol. Pharmacol.* 17, 14–23

12 Pfeuffer, T. (1979) *FEBS Lett.* 101, 85–89

13 Limbird, L. E., Gill, D. M. and Lefkowitz, R. J. (1980) *Proc. Natl. Acad. Sci. U.S.A.* 77, 775–779

14 Cassel, D., Levkovitz, H. and Selinger, F. (1977) *J. Cyclic Nucleotide Res.* 3, 393–406

15 DeLean, A., Stadel, J. and Lefkowitz, R. J. *J. Biol. Chem.* (in press)

16 Su, Y. F., Cubeddu, X. L. and Perkins, J. P. (1976) *J. Cyclic Nucleotide Res.* 2, 257–270

17 Caron, M. G., Srinivasan, Y., Pitha, J., Kociolek, K. and Lefkowitz, R. J. (1979) *J. Biol. Chem.* 254, 2923–2927

18 Vanquelin, G., Geynet, P., Hanoune, J. and Strasberg, A. D. (1977) *Proc. Natl. Acad. Sci. U.S.A.* 74, 3710–3714

Identification and significance of beta-adrenoceptor subtypes

Stefan R. Nahorski

Department of Pharmacology and Therapeutics, Medical Sciences Building, University of Leicester, University Road, Leicester LE1 7RH, U.K.

At the present time we have little reason to doubt Ahlquist's original proposal that catecholamines exert their physiological effects by interacting with alpha or beta adrenoceptors. However, more recent developments have suggested that these receptors may not be homogeneous and the existence of alpha$_1$, alpha$_2$, beta$_1$ and beta$_2$ adrenoceptors has received considerable experimental support.

The possibility that beta-adrenoceptors do not form a homogeneous population was first suggested by Moran[1] when he drew attention to the fact that beta-adrenoceptor antagonists possessing alkyl substituents on the alpha-carbon of the ethanolamine side chain display different potencies in blocking vascular and cardiac responses to catecholamines. However, it was Lands et al.[2] who first suggested the subclassification of beta-adrenoceptors into beta$_1$ and beta$_2$. This division was made on observations that the rank order of potency of catecholamine agonists fell into two distinct categories depending upon the tissue response examined. The receptors in heart and adipose tissue were classified as beta$_1$, whereas those in bronchi and vascular smooth muscle were designated beta$_2$. Further support for this sub-

classification of beta-adrenoceptors has come from subsequent observations that certain antagonists (e.g. practolol, atenolol, metoprolol) are more effective in blocking catecholamine responses in the heart, whereas certain agonists such as salbutamol and soterenol were clearly more selective for respiratory tissue. However, this organ-specific dual receptor hypothesis has not been universally accepted since several authors have reported a spectrum of apparent affinities for agonists and antagonists between tissues of various species (for an excellent review of these pharmacological studies see Ref. 3). It is clear however, that the precise analysis of drug-receptor interactions using classical pharmacological approaches in whole tissue can be prone to a number of serious limitations. Thus with agonists there could be quite different rates of loss by uptake or non-receptor binding between various tissues. An apparent selectivity for an agonist could also be manifest by a different 'receptor reserve' between tissues. Even with antagonists, the distribution of the compound both between and within a tissue could create an apparent selectivity related to the physico-chemical nature of the drug and be quite independent of a selective interaction with the receptor. In this respect, it should be noted, for example, that almost all beta₁-selective antagonists are more hydrophilic than non-selective antagonists. The proposal by Carlsson and his colleagues[4] that both beta₁ and beta₂ adrenoceptors may co-exist in the same organ and under certain circumstances may mediate the same physiological response, could also influence the analysis of pharmacological responses of whole organs.

In order to overcome the inherent problems associated with drug-receptor interactions in intact pharmacological preparations, several research groups have recently used radiolabelled ligand binding techniques to characterise and quantify adrenergic receptors. It is probably because of these new approaches that a clearer picture of the nature of beta-adrenoceptor subtypes has emerged.

Identification of beta-adrenoceptor subtypes using radioligand techniques

Over the last few years there have been quite dramatic developments in beta-adrenoceptor research which have been directly related to the use of direct receptor labelling techniques and the recognition that adenylate cyclase stimulation is one of the primary events that follows occupation of the receptor by an agonist[5]. The inherent advantage of these techniques for receptor characterisation lies with the use of membrane preparations in which problems of drug-receptor equilibrium can be minimized. Moreover, the direct labelling of receptors allows a precise quantification of receptor subtype density that is not complicated by events occurring 'down stream' of the receptor itself. On the other hand, it should be appreciated that the lack of a final biological response can be a distinct disadvantage in attempts to relate receptor heterogeneity to functionally different cell types.

The most commonly used ligands in this field are the non-selective beta-adrenoceptor antagonists (−)-[³H]dihydroalprenolol ([³H]DHA) and (±)-[¹²⁵I]iodohydroxybenzylpindolol ([¹²⁵I]IHYP). Although the latter ligand possesses the advantage of a considerably higher specific activity, this can be outweighed by higher non-specific binding and an apparent association with 5-HT recognition sites in some tissues. For this reason, the majority of our studies have been performed with [³H]DHA.

The first indication that the direct labelling of beta-adrenoceptors might reveal new information concerning the existence of receptor subtypes came from studies in rat cerebral cortex and lung[6,7]. In these experiments [³H]DHA could be displaced by other non-selective antagonists from membranes of either tissue in a manner suggesting interaction with a homogeneous receptor population with binding governed by the law of mass action. However, certain drugs that have been reported to display beta₁ selectivity in intact preparations generated biphasic displacement curves in both tissues. Our interpretation of these data was that cortex and lung possess *both* beta₁ and beta₂ adrenoceptors and that the selective antagonists possess markedly different affinities for the two populations of receptor subtypes that were being labelled by [³H]DHA.

Initially an Eadie-Hofstee transformation of the data allowed an estimation of the proportion of receptor subtypes and the relative affinities of the competing drug. Using these techniques it could be shown that in cerebral cortex about 65% of the receptors possessed high affinity for beta₁ antagonists, whereas in lung only about 20% of the sites were displaced with high affinity by these beta₁-selective agents. It was gratifying that despite different proportions of high and low affinity sites in lung and brain, the affinity constants for beta₁ selective agents for each particular subpopulation of sites were identical for the two tissues. However, conclusive evidence that these sites represented beta₁ and beta₂ adrenoceptors only came when it could be shown that highly selective beta₂ adrenoceptor drugs such as procaterol and ICI 118.551 generated curves that were the precise 'mirror image' of those seen with beta₁ selective drugs. The curves generated by the beta₁-selective

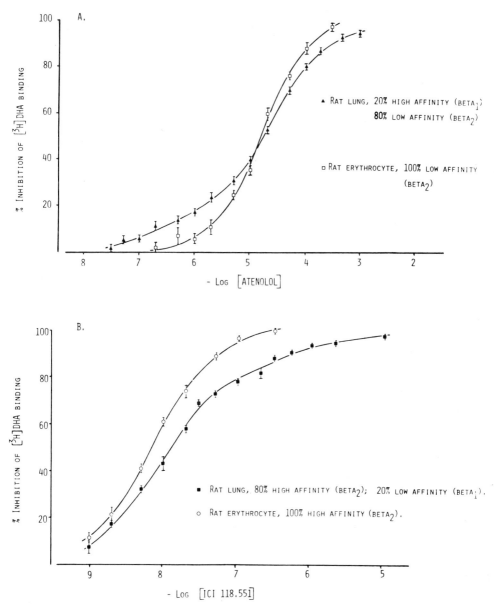

Fig. 1. Inhibition of [³H]DHA binding to rat lung and erythrocyte membranes by (A) the beta₁ selective antagonist atenolol and (B) the beta₂ selective antagonist ICI 118.551 [erythro-1-(7-methylindan-4-yloxy-3-isopropyl-aminobutan-2-ol)]. Computer-assisted curve fitting revealed that the erythrocyte membranes possess a homogeneous population of sites that have low affinity for atenolol and high affinity for ICI 118.551. On the other hand, the lung curves are best fitted by a two-site model in which 20% of the sites have high affinity for atenolol and the remaining 80% high affinity for ICI 118.551.

antagonist atenolol and the beta$_2$-selective antagonist ICI 118.551 in rat lung and erythrocyte are illustrated in Fig. 1. It could also be demonstrated that the prior selective occupation of only one class of receptor subtype converted the biphasic displacement curves of agents that were selective for the other subtype into uniphasic mass action curves. This suggests that beta$_1$ and beta$_2$ adrenoceptors can co-exist in tissues in a non-interacting manner and are probably quite separate entities.

Subsequent to these initial studies, it has been correctly pointed out that the linear transformation methods of analysing the complex curves generated by beta$_1$ or beta$_2$ selective agents can be inaccurate, particularly with drugs that display less than 30 fold selectivity for the receptor subtypes. The advantages of using non-linear curve fitting techniques have been detailed recently[5] and this form of analysis is routinely used in our laboratory at the present time. However, in view of the fact that only highly selective agents were used in our own studies, direct comparison of

linear and non-linear handling of the data has yielded virtually identical results.

Using these techniques, analysis of the proportions of beta$_1$ and beta$_2$ adrenoceptors can be made in a variety of mammalian tissues and a summary of studies using [^{3}H]DHA are shown in Table I. These results suggest that (a) there are only two beta-adrenoceptor subtypes in mammalian tissues and that they correspond to the beta$_1$ and beta$_2$ subclasses suggested previously from classical pharmacological approaches, and (b) in most tissues both beta$_1$ and beta$_2$ adrenoceptors co-exist but the proportion of each subtype depends upon the tissue being examined. Not only are there differences between various tissues of the same species, but even in the same tissue of different species. Thus, unlike other mammalian species, beta$_1$ rather than beta$_2$ receptors predominate in rabbit lung[8].

More recently, using the radioligand [^{125}I]IHYP and techniques virtually identical to those above, Minneman and his colleagues have been able to confirm the co-

TABLE I. Tissue distribution of beta-adrenoceptor subtypes

Tissue	Total beta-adrenoceptor density (fmol × mg prot.$^{-1}$)	% Beta$_1$	Beta$_2$
Rat lung	400	20	80
Rabbit lung	350	80	20
Bovine lung	250	25	75
Rat ventricle	50	65	35
Rat spleen	250	35	65
Rat uterus (oestrogen dominated)	100	20	80
Rat uterus (progesterone dominated)	100	0	100
Rat erythrocyte	100	0	100
Rat reticulocyte	600	0	100
Rat cerebral cortex	120	65	35
Rat cerebellum	50	0	100
Rat striatum	100	65	35
Rat limbic forebrain	70	55	45

All studies were performed using the ligand (−)-[^{3}H]dihydroalprenolol. Proportions of beta$_1$ and beta$_2$ sites were estimated by computer-assisted curve fitting of the atypical displacement curves generated by highly selective beta$_1$ or beta$_2$ agents (see text).

presence of beta₁ and beta₂ adrenoceptors in several tissues[9] Overall, therefore, the use of receptor labelling techniques has provided direct evidence for Carlsson's original suggestion that beta₁ and beta₂ adrenoceptors can co-exist in the same tissue. These findings may provide some explanation for data from classical pharmacological approaches that could not be easily reconciled with a simple organ-specific dual beta-adrenoceptor concept. Thus, if most tissues possess both receptor subtypes, and if (and this is a critical assumption) both populations mediate the same physiological function, then the affinity of an antagonist or the potency of an agonist will depend in part on the relative proportion of each subtype in the tissue in question.

Cellular localization of beta-adrenoceptor subtypes

It is far from clear whether we can extend the findings using ligand binding studies described above to conclude that both subtypes co-exist on the same *cell*. It must be stressed that the tissues in which both beta-receptor subtypes apparently co-exist possess several different cell types. Thus, the presence of both subtypes within a tissue may merely reflect this cellular heterogeneity. A recent examination of membranes prepared from rat erythrocytes and reticulocytes has revealed that these homogeneous populations of cells only possess beta₂ adrenoceptors. On the other hand, investigation of the ontogeny of rat lung beta-adrenoceptors indicated that the proportion of beta₁ and beta₂ sites remained constant, despite the markedly different ratio of cell types found at various stages of development. This could indicate that both receptor subtypes are present on the same cell type and would support pharmacological evidence that catecholamines mediate smooth muscle relaxation in guinea-pig trachea by interacting with both beta₁ and beta₂ adrenoceptors[10].

An attractive concept is that beta-adrenoceptor heterogeneity could reflect a differential pre- and post-synaptic distribution of beta₁ and beta₂ receptors. Since pre-synaptic beta-adrenoceptors may be involved in the regulation of noradrenaline release, the therapeutic consequences of such a phenomenon could be particularly important. However, chemical sympathectomy with 6-hydroxydopamine has failed to indicate any significant proportion of presynaptic beta receptors in spleen, and has revealed that both beta₁ and beta₂ subtypes co-exist postsynaptically in this tissue[11].

In the central nervous system it is probable that glial cells and possibly cerebral blood vessels as well as neurones, possess beta-adrenoceptors. In support of this it has been recently shown that specific lesions of nerve cell bodies in the striatum using the neurotoxin kainic acid leads to only a small (30%) loss of beta-adrenoceptor binding sites[12]. However, it is of interest that the sites lost are almost exclusively from the beta₁ population. A better indication of the precise cellular localization of beta₁ and beta₂ adrenoceptors is clearly required. It seems probable that refinement of the autoradiographic studies available now for several neurotransmitter receptors may provide better clues in the near future.

The question of selective beta-adrenoceptor agonists

One of the corner-stones of the original suggestion of beta-adrenoceptor heterogeneity is that certain agonists such as salbutamol, soterenol and terbutaline are clearly selective in their ability to produce bronchodilation rather than cardiac stimulation. This property forms the basis of their therapeutic usefulness. It is therefore particularly surprising to find that these agents do not demonstrate any significant selective *affinity* for beta₂ adrenoceptors when examined in ligand binding studies[6,8,9]. Perhaps a partial explanation for this paradox lies in the finding that these agonists may have a greater *efficacy* at beta$_2$ adrenoceptors than at beta$_1$ sites. Thus, in rat and rabbit lung that possess a predominance of beta$_1$ and beta$_2$ receptors respectively, although salbutamol and soterenol displace [^{3}H]DHA binding with equal affinity in both preparations, these agonists stimulate adenylate cyclase to a greater extent in beta$_2$-rich rat lung membranes[13]. If one considers the large amplification of the signal from receptor to the final biological response, these findings could provide an answer to the tissue selectivity of these beta$_2$ agonists. Several years ago, Jenkinson[14] proposed a related explanation for the selectivity of salbutamol-like compounds. He argued that if there was a larger 'receptor reserve' for pulmonary over cardiac beta-adrenoceptors, then partial agonists such as salbutamol could display tissue selectivity without possessing a selective affinity for beta$_2$ adrenoceptors. In any case, the demonstration that certain agonists apparently bind to beta$_1$ and beta$_2$ adrenoceptors with equal affinity but induce a more efficient coupling of beta$_2$ adrenoceptors to adenylate cyclase, could suggest that there is a different interaction between the two beta-adrenoceptor recognition sites with guanine nucleotide binding proteins and/or catalytic sites of adenylate cyclase. Preliminary experiments in this laboratory also suggest that there may be a different influence of guanine nucleotides on the interaction of agonists (see Ref. 5) with beta$_1$ and beta$_2$ adrenoceptors in lung, perhaps again suggesting that there are different receptor–effector coupling relationships between the two subtypes.

Atypical beta-adrenoceptor subtypes

As discussed above, information from ligand binding studies suggests that mammalian tissues contain only two beta-adrenoceptor subtypes, since the pharmacological properties of beta$_1$ or beta$_2$ binding sites are virtually identical in all mammalian tissues examined to date. However, the recent extension of these techniques to non-mammalian tissues has revealed significant differences in the characteristics of the beta-adrenoceptors, particularly in certain non-mammalian erythrocytes. Avian and frog erythrocytes have always been considered to possess beta$_1$ and beta$_2$ adrenoceptors respectively, and indeed have been used as model systems for these receptor subtypes. Minneman and colleagues first suggested that the affinity of certain beta-adrenergic agents for the turkey erythrocyte beta-adrenoceptor differed significantly from either the beta$_1$ or beta$_2$ mammalian receptors[15]. We have recently also noted a similar atypical site on chick erythrocyte, and have further shown that the beta-adrenoceptor on frog erythrocyte is also different to mammalian beta$_1$ or beta$_2$ receptors. Since the differences in these erythrocyte preparations are particularly evident in respect to the affinities of certain beta-adrenoceptor antagonists but the rank order of catecholamines still resembles beta$_1$ or beta$_2$ receptors, it is debatable whether one should assign a new classification to these beta-adrenoceptors. However, in view of other differences relating to receptor–adenylate cyclase coupling, one should be cautious in any extrapolation of results from these model systems to the situation at mammalian beta$_1$ and beta$_2$ adrenoceptors.

Conclusion

It is quite evident that the use of ligand binding has provided new information concerning the precise pharmacological characterization and quantification of beta-adrenoceptor subtypes. This biochemical approach suggests strongly that, in mammalian tissues at least, there are only two subtypes that correspond pharmacologically to beta$_1$ and beta$_2$ subclasses, and further that most tissues possess *both* subtypes.

Although it can be shown that beta₁ and beta₂ adrenoceptors can co-exist in tissues in a non-interacting manner, the definitive identification of distinct receptor subtypes requires the physical separation of the subtypes from solubilised membrane preparations. It is possible that a single beta-adrenoceptor protein could display beta₁ or beta₂ characteristics, depending upon which accessory binding sites are revealed in the intact membranes. We feel this is unlikely in view of preliminary studies from this laboratory which indicate that solubilised preparations, free of the restraint of the membranes, still retain beta₁ and beta₂ pharmacological characteristics. Moreover, the individual recovery of the subtypes from membranes containing both subclasses is very different and it seems probable that the receptor subtypes are separate proteins. Conclusive evidence on this point, however, awaits the physical separation of beta₁ and beta₂ adrenoceptors from solubilized preparations.

There has been considerable debate concerning the 'physiological' or 'pharmacological' significance of beta₁ and beta₂ adrenoceptors and the relative roles of adrenaline and noradrenaline in the initiation of responses at these sites[3]. It seems probable that there will soon be new information on the individual receptor-effector coupling characteristics and precise localisation of beta₁ and beta₂ subtypes in tissues. At that time we should be in a better position to discuss the relative importance of these receptors in normal and diseased tissues.

Acknowledgements

It is a pleasure to acknowledge the important contributions of a number of colleagues – D. Barnett, K. Dickinson, A. Richardson, E. Rugg and A. Willcocks – to different aspects of the work described in this review.

Reading list

1 Moran, N. C. (1966) *Pharmacol. Rev.* 18, 503–512

2 Lands, A. M., Arnold, A., McAuliff, J. P., Laduena, F. P. and Brown, T. G. (1967) *Nature (London)* 214, 597–598

3 Daly, M. J. and Levy, G. P. (1979) in *Trends in Autonomic Pharmacology* (Kalsner, D., ed.), Vol. 1, pp. 347–385, Urban and Schwarzenberg, Baltimore and Munich

4 Carlsson, E., Ablad, B., Brandstrom, A. and Carlsson, B. (1972) *Life Sci.* 11, 953–958

5 Lefkowitz, R. J. and Hoffman, B. B. (1980) *Trends Pharmacol. Sci.* 1, 314–318

6 Nahorski, S. R. (1978) *Eur. J. Pharmacol.* 51, 199–209

7 Barnett, D. B., Rugg, E. and Nahorski, S. R. (1978) *Nature (London)* 273, 166–168

8 Rugg, E., Barnett, D. B. and Nahorski, S. R. (1978) *Mol. Pharmacol.* 14, 996–1005

9 Minneman, K. P., Hegstrand, L. R. and Molinoff, P. B. (1979) *Mol. Pharmacol.* 16, 34–46

10 Zaagsma, J., Oudhof, R., van der Heijden, P. J. and Plantje, J. F. (1979) in *Catecholamines, Basic and Clinical Frontiers* (Usdin, E., Kopin, I. J. and Barchas, J., eds), Vol. 1, pp. 435–437

11 Nahorski, S. R., Barnett, D. B., Howlett, D. R. and Rugg, E. (1979) *Naunyn-Schmiedebergs Arch. Pharmacol.* 307, 227–233

12 Nahorski, S. R., Howlett, D. R. and Redgrave, P. (1979) *Eur. J. Pharmacol.* 60, 249–252

13 Nahorski, S. R. (1979) in *Medicinal Chemistry VI* (Simkins, M., ed.), pp. 147–156, Cotswold Press Limited, Oxford

14 Jenkinson, D. H. (1973) *Br. Med. Bull.* 29, 142–147

15 Minneman, K. P., Weiland, G. A. and Molinoff, P. B. (1980) *Mol. Pharmacol.* 17, 1–7

β-Adrenergic receptor subtypes

G. Leclerc, B. Rouot, J. Velly and J. Schwartz

Institut de Pharmacologie (ERA 142 CNRS FRA 6 INSERM), 11 rue Humann, 67000 Strasbourg, France.

In 1948, Ahlquist[1] proposed dividing the adrenergic receptors into α- and β-receptors according to their responses to phenylephrine and isoproterenol. From 1958 onwards, broad series of β-agents, both agonists and antagonists, were synthesized. Then Lands[2] classified the β-receptors in two autonomous subgroups, β_1 (cardiac) and β_2 (bronchial and vascular). This article deals with the legitimacy of that classification.

β-stimulants

Most of these are isoproterenol-related. The importance of the catechol function and particularly the major role of the meta-hydroxyl have been stressed. This hydroxyl can however by replaced with an amine, an alkylamine or even, without any loss of performance, by a CH_3SO_2NH-(soterenol), or $-CH_2OH$ (salbutamol) group. The carbostyril derivatives, related to soterenol, are particularly active[3]; some of them seem to have remarkable β_2-selectivity, but this needs to be confirmed. Curiously, one of the least selective of these compounds, procaterol, is the only one freely available. The partial β-agonist activity of the halogen derivatives such as clenbuterol and especially clorprenaline, whose aromatic nucleus is devoid of groups which might be protonated or which might form a hydrogen bond is surprising. It should be noted that while the L-isomer of clenbuterol is a α-agonist, its D-isomer is a β-blocking agent.

The asymmetric carbon of the ethanolamine must necessarily have the (R)-configuration, since the (S)-isomer of isoproterenol is practically inactive. Substitution of the α carbon of the side chain of isoproterenol by an ethyl group gives a β_2-stimulant (isoetarine) of which only the erythro-isomer is active. However, the β-stimulant activity is not necessarily linked to the presence of an ethanolamine chain, since the structural analog of isoproterenol with an oxypropanolamine chain is highly active[4]: the same applies to tazolol which, in addition to this same side chain has, as an aromatic nucleus, a non-substituted thiazole ring. Finally, while substitution on nitrogen by t-Bu or i-Pr groups is known to favor a β_2-action, certain 'heavy' amines have improved cardioselectivity.

Pharmacological experiments with these molecules are subject to criticism insofar as most authors take isoproterenol as a reference for the activities of β-agonists. Now, it is by no means sure that this molecule is equipotent for β_1 and β_2-receptors[4]. Moreover, in binding studies, with [³H]dihydroalprenolol [[³H]DHA] as ligand, we found that (±)-isoproterenol has a β_1/β_2 selectivity ratio of about 30. These results are comparable with those of U'Prichard[5] who found a β_1/β_2 ratio of about 10 for (−)-isoproterenol. Nevertheless Minneman[6,7], using iodohydroxybenzylpindolol (IHYP) as ligand, found a ratio close to 2. Computerized graphic analysis of the curvilinear Hofstee plots, resulted in a ratio of about 1. Furthermore, our binding technique[8] (Table I) did not bear out the alleged β_2-selectivity of salbutamol, terbutaline, fenoterol and soterenol. Only procaterol shows some selectivity. Our results have been confirmed and an attempt has been made to explain the β_2-selectivity of salbutamol on the basis of its inability to stimulate cardiac adenylate cyclase[6]. It should be noted that with terbutaline, the action on adenylate cyclase is similar in the heart and lung. The technique for measuring adenylate cyclase activity does not adequately reflect the pharmacological activity. Thus, Minneman *et al.* did not observe any cardiac adenylate cyclase stimulation under the effect of zinterol, salmefamol, salbutamol or soterenol, whereas these drugs definitely act on the heart, even though they may have a certain β_2-selectivity. However, Minneman[7] claims some selectivity for zinterol and salmefamol. In fact, we consider a possible difference in the structures of β_1 and β_2-receptors and the existence of subgroups as hypothetical, to say the least. It has been suggested[9] that the chronotropic or inotropic action of these molecules might be due to different receptors, and that the vascular receptors differ from the tracheal ones. These conclusions are also subject to doubt, since once again, isoproterenol was taken as a non-selective reference substance.

β-Blockers

These are very homogeneous in their structures and stem either from arylethanolamines (dichloroisoproterenol, pronethalol, sotalol) or from aryloxy-propanolamines for the more recent ones (propranolol, practolol, pindolol). We have shown that slightly different structures can also be highly active, e.g. IPS 339 [(tertiobutyl-amino-3-ol-2-propyl) I oximo 9-fluorene] which is a highly selective β_2-blocker[10]. Acetophenone derivatives have also proved interesting, and most of them are β_2-selective[11]. Finally, we have recently synthesized β-blockers mostly lacking aromatic nuclei; they are aliphatic oximes or aliphatic aminoalkylethers[12]. Thus, contrary to general opinion, the aromatic nucleus is probably not quite as important as it was thought to be.

In a pharmacological study of 6 β-blockers[13] (propranolol, pindolol, practolol, atenolol, metoprolol and IPS 339), the β_1/β_2-activity ratios were identical for each molecule, *in vitro* in the isolated organ as well as *in vivo* in the guinea-pig and the dog. Selectivity does not vary from one animal species to another and remains identical whether we take the chronotropic or inotropic action as a model for β_1-receptors, or the trachea or blood vessels for β_2-receptors.

The selectivities obtained by these traditional pharmacological techniques are in close agreement with the inhibition constants measured in lung and heart preparations by binding. So, in our binding assays

TABLE I

β-agonists	Heart K_i (μM)	Isolated atria pD$_2$	Lung K_i (μM)	Isolated trachea pD$_2$	β_2/β_1 activity	
					Antilog ΔpD$_2$*	$\dfrac{1/K_i \text{ lung}}{1/K_i \text{ heart}}$
(±) Salbutamol	1.9	5.9	3.0	7.13	17	0.63
(±) Terbutaline	9.4	5.13	9.2	6.0	7.4	1.02
(±) Procaterol	5.6	7.43	1.2	8.46	10.5	4.7
(±) Fenoterol	2.3	6.75	1.7	7.75	10	1.35
(±) Soterenol	1.2	6.25	2.9	7.72	29.5	0.4

* Antilog ΔpD$_2$ = pD$_2$ on trachea – pD$_2$ on atria.
Binding assays were made using iodohydroxybenzylpindolol (IHYP) as ligand.

(Table II), acebutolol, atenolol and prac-tolol, which are respectively, 3, 4 and 9 times more active on the heart than on the lung, are β_1-selective. Sotalol and butox-amine, which are respectively 7 and 5 times more active on the lung, are β_2-selective. The so-called 'non-specific' β-blockers, such as propranolol and oxprenolol do in fact have similar inhibition constants for heart and lung. But selectivities obtained with membrane preparations are always lower than those observed *in vivo* and with isolated organs (*in vivo*). The same is not

true for pindolol, which we found to have a β_1-selectivity not observed by Minneman who used IHYP as ligand. The partial agonist action of pindolol might account for this discrepancy. What is in fact surpris-ing is that, despite the synthesis of thousands of β-blockers, in pursuit of a highly selective compound, the tests car-ried out on membrane preparations now show that the most selective is still prac-tolol, which has a β_1/β_2 affinity ratio of only 9.

Our present data in fact shed doubts on

TABLE II

β-Antagonists	Heart K_i (nM)	Isolated atria pA2	Lung K_i (nM)	Isolated trachea pA2	β_1/β_2 activity Antilog ΔpA2*	$1/K_i$ heart / $1/K_i$ lung
($\pm$)-Propranolol	2.4	8.62	2.7	8.47	1.4	1.12
($\pm$)-Oxprenolol	2.6	8.73	1.7	8.46	1.9	1.12
($\pm$)-Pindolol	2.9	9.19	90	8.97	1.66	31
($\pm$)-Alprenolol	9	8.5	3	8.43	1.2	0.33
($\pm$)-Labetalol	27	8.12	22.5	7.89	1.7	0.83
($\pm$)-Acebutolol	1200	6.84	3700	5.47	23.4	3
($\pm$)-Atenolol	1400	7.66	6250	6.13	33.9	4.46
($\pm$)-Sotalol	2400	6.40	340	6.47	0.85	0.14
($\pm$)-Practolol	2500	6.85	22000	5.13	52.5	8.8
($\pm$)-Butoxamine	4200	5.24	750	6.44	0.08	0.18

* Antilog ΔpA2 = pA2 on atria – pA2 on trachea.
Binding assays were made using [^{3}H]dihydroalprenolol as ligand.

TABLE III

		β_1 activity		β_2 activity		β_1/β_2 activity	
Practolol	Para	6.98*	6.12†	5.14*	5.18†	69*	8.7†
	Ortho	6.56*	5.23†	6.82*	5.14†	0.55*	1.2†
Oxprenolol	Para	8.39*	6.5†	6.80*	–	40*	–
	Ortho	9.44*	6.5†	8.80*	7.16†	4.4*	0.22†
Alprenolol	Para	8.09*	–	6.05*	–	110*	–
	Ortho	8.41*	–	9.14*	–	0.2*	

* Our pA2 values were obtained from guinea-pig auricle (β_1) and trachea (β_2).
† Chronotropic action (β_1) and peripheral resistance (β_2) on the dog *in vivo*, according to Vaughan Williams.

the existence of β-subgroups. There are some authors who envisage the constant association within a given organ of β_1 and β_2-receptors. So it is that Ablad and Carlsson[14] using human heart fragments, stimulated with noradrenaline or adrenaline observed that propranolol, a non-specific β-blocker, inhibits both catecholamines identically. On the other hand, metoprolol, a β_1-blocker, is more effective in inhibiting the action of noradrenaline than that of adrenaline. But this work can be criticized since selectivity of metoprolol is not evident. In view of its clear β_2-selectivity, we made use of IPS 339, showing in guinea pig heart, with this β-blocker, that the shift in the dose-response curve is identical, whether adrenaline or noradrenaline is used as agonist. Our results are therefore opposed to those of Ablad and Carlsson. Nahorski[15] using [³H]dihydroalprenolol as ligand has observed that on rat lung the order of affinity of catecholamines is: isoproterenol > adrenaline > noradrenaline and in rabbit, isoproterenol > adrenaline = noradrenaline. On the basis of Land's classification, one might conclude that in the rat lung, the predominant receptors are β_2, and in the rabbit β_1. These data are con-

firmed by Scatchard analysis of the displacement curves. It should however be noted that for each molecule, the differences in affinity for β_1 and β_2-receptors are very small. In fact, numerous reservations can be made about the use of binding techniques: it is by no means sure that displacement of β-blocker, in this case [³H]dihydroalprenolol, is valid in studying β-agonists such as salbutamol and terbutaline. U'Prichard[5] in fact has shown that determining K_i for salbutamol and terbutaline by the displacement of [³H]adrenaline gives lower values than with [³H]dihydroalprenolol. It should however be noted that experiments with [³H]adrenaline require quite high concentrations of pyrocatechol and phentolamine in the incubation medium. The presence of these substances may well interfere with specific binding. Then again, electron microscopy reveals heterogeneity in the membrane preparations described in the literature used in such studies and casts serious doubts on the validity of some published results.

Our work, then, lends weight to the theory that selectivity which is always low (except for a few molecules, among them,

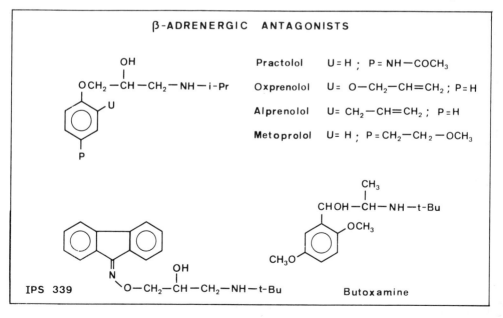

β-ADRENERGIC ANTAGONISTS

Practolol   U = H ; P = NH—COCH₃

Oxprenolol   U = O—CH₂—CH=CH₂ ; P = H

Alprenolol   U = CH₂—CH=CH₂ ; P = H

Metoprolol   U = H ; P = CH₂—CH₂—OCH₃

IPS 339

Butoxamine

practolol in particular) does not depend on the type of receptor, but on factors controlling bio-availability; the physical properties of the molecule seem more important than its chemical nature. This would tie up with the views of Ariëns[16] on blocking agents, but would seem somewhat surprising for agonists, which do not operate on secondary lipophilic sites. Vaughan Williams[17] stressed steric factors, showing that molecules whose aromatic nucleus is para-substituted are more cardioselective than the ortho-substituted derivatives since β_2-receptors cannot accommodate a group in the para-position. Thus, Vaughan Williams, working on the dog *in vivo*, found that para-practolol and para-oxprenolol are cardioselective, whereas the ortho-substituted analogs are not (Table III). These results have been confirmed by Ablad and Carlson, working on isolated organs. To some extent, we also found this selectivity for the para-substituted derivatives *in vitro*, but this was not clearly borne out by binding techniques (Table IV).

In fact, the debate remains open. What must be done is to see if specific β_1 or β_2 drugs can be developed. It must also be

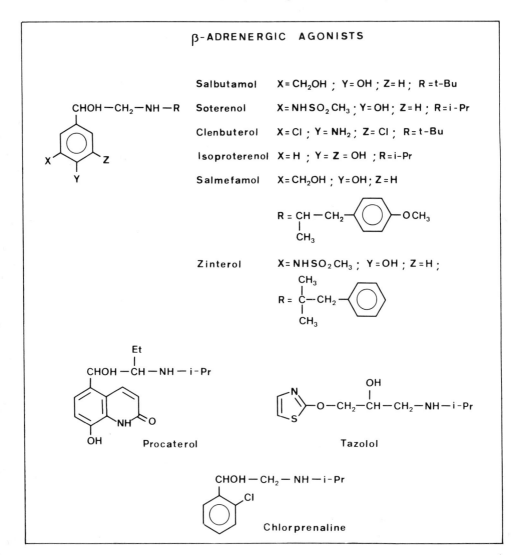

TABLE IV

		Heart K_i (nM)	Lung K_i (nM)	β_1/β_2 activity
Practolol	Para	500	5000	10
	Ortho	225	1385	6
Oxprenolol	Para	205	615	3
	Ortho	2.6	1.7	0.6
Alprenolol	Para	330	923	2.8
	Ortho	9	1.8	0.2

K_i values obtained from the inhibition of [^{3}H]dihydroalprenolol binding on heart and lung membrane preparations.

determined whether bio-availability is a decisive factor, and if so, more appropriate screening techniques will have to be evolved. Further, the factors governing the coupling between β-receptors and adenylate cyclase must be more fully characterized in mammalian systems.

Reading list

1 Ahlquist, R. P. (1948) *Am. J. Physiol.* 153, 586–600
2 Lands, A. M., Luduena, F. P. and Buzzo, H. J. (1967) *Life Sci.* 6, 2241–2249
3 Yoshizaki, S., Tanimura, K., Tamada, S., Yabuuchi, Y. and Nakagawa, K. (1976) *J. Med. Chem.* 19, 1138–1142
4 Kayser, C. (1979) in *Advances in Receptor Chemistry* (Gualtieri *et al.*, ed.), pp. 319, Elsevier/North-Holland Biomedical Press, Amsterdam
5 U'Prichard, D. C., Bylund, D. B. and Snyder, S. H. (1978) *J. Biol. Chem.* 253, 5090–5102
6 Minneman, K. P., Hegstrand, L. R. and Molinoff, P. B. (1979) *Mol. Pharmacol.* 16, 21–33
7 Minneman, K. P., Hedberg, A. and Molinoff, P. B. (1979) *J. Pharmacol. Exp. Ther.* 211, 502–508
8 Bieth, N., Rouot, B., Schwartz, J. and Velly, J. (1980) *Br. J. Pharmacol.* 68, 563–569
9 O'Donnell, S. R. and Wanstall, J. C. (1976) *Br. J. Pharmacol.* 57, 369–373
10 Imbs, J. L., Miesch, F., Schwartz, J., Velly, J., Leclerc, G., Mann, A. and Wermuth, C. G. (1977) *Br. J. Pharmacol.* 60, 357–362
11 Leclerc, G., Mann, A., Wermuth, C. G., Bieth, N. and Schwartz, J. (1977) *J. Med. Chem.* 20, 1657–1662
12 Leclerc, G., Bieth, N. and Schwartz, J. (1980) *J. Med. Chem.* 23, 620–624
13 Miesch, F., Bieth, N., Leclerc, G. and Schwartz, J. (1978) *J. Pharmacol.* 9(4), 297–308
14 Ablad, B., Carlsson, B., Carlsson, E., Dahlof, C., Ek, L. and Hultberg, E. (1974) *Adv. Cardiol.* 12, 290–302
15 Nahorski, S. R. (1978) *Eur. J. Pharmacol.* 51, 199–209
16 Ariëns, E. J. (1979) *Trends Pharmacol. Sci.* 1, 11–15
17 Vaughan Williams, E. M., Bagwell, E. E. and Singh, B. N. (1973) *Cardiovascular Res.* 7, 226–240

New directions in adrenergic receptor research

Part II

Robert J. Lefkowitz and Brian B. Hoffman

Howard Hughes Medical Institute Laboratory, Departments of Medicine (Cardiovascular) and Biochemistry, Duke University Medical Center, Durham, North Carolina 27710, U.S.A.

The first successful direct identification of alpha-adrenergic receptors with radioligands in 1976[1,2] came shortly after similar studies had accomplished with beta-adrenergic receptors. Over the past several years a great number of radioligands have been developed which appear to bind to alpha-adrenergic receptors. These ligands are listed in Table I. As opposed to the beta-adrenergic receptors, for which all of the antagonist ligands appear to be non-subtype selective (i.e. equal affinity for $beta_1$ and $beta_2$ receptors), the alpha-adrenergic receptors have a variety of subtype-selective radioligands. In addition, there are further complexities in interpreting the patterns of radioligand binding because apparently $alpha_2$, but not $alpha_1$, adrenergic receptors are regulated by guanine nucleotides[3]. The large number of ligands available and the complexities introduced by the existence of alpha receptor subtypes and the effects of guanine nucleotides on agonist binding has made the interpretation of the recent literature difficult.

Before discussing the labelling of alpha-adrenergic receptors by various radioligands, it is necessary to review the classification of alpha-adrenergic receptors into two subtypes. The physiological data supporting the concept that there are discrete alpha-adrenergic receptor subtypes have been developed over the past ten years and have been extensively reviewed[4,5]. Unfortunately, there has not been uniformity in the terminology used to describe these subtypes. The autoregulatory alpha receptors that inhibit norepinephrine release from nerve terminals have been called both presynaptic alpha receptors and $alpha_2$ receptors whereas the alpha receptors typically found on effector cells have been termed postsynaptic alpha receptors and $alpha_1$ receptors. The classification based on the terms $alpha_1$ and $alpha_2$ seems to be preferable[6]. The alpha receptor subtypes may be demarcated by their pharmacological properties in response to a variety of drugs. Thus far, no simple, generally agreed upon definition of $alpha_1$ and $alpha_2$ receptors in terms of a potency series of agonist drugs has emerged analogous to that for the beta receptor subtypes. Agonists such as clonidine, methoxamine and phenylephrine and others have been proposed to fill such a role in demarcating the alpha receptor subtypes in physiological studies. However, there has been great variability in the apparent selectivity of these drugs from one animal model to another; a drug might be apparently quite selective in one system and not so in another. This could reflect real differences in their ability to interact with the alpha receptors, suggesting that the classification of alpha receptor subtypes into just two subtypes is oversimplified. Alternatively, since such physiological studies provide relatively indirect estimates of receptor occupancy, it is possible that other factors may also be involved. These include differences in drug diffusion, metabolism in tissues, drug uptake, agonist intrinsic activity, receptor-response coupling and 'spare receptors' as examples. All of these could play a role in confounding the interpretation of experiments designed to measure drug affinity based on physiological responses in comparison with affinities determined by radioligand binding. As noted in Part I, amongst the antagonists, prazosin has most often been found to be highly alpha₁ selective. Yohimbine is generally alpha₂ selective although its relative selectivity appears not to be as great as the alpha₁ selectivity of prazosin.

Radioligand binding studies have been used to delineate the alpha-adrenergic receptor subtypes. Recently two different approaches to this problem have been described. One approach is to use a non-subtype selective radioligand to label the entire alpha receptor population of the tissue and then to determine the proportion of alpha₁ and alpha₂ receptors by constructing competition curves with subtype selective drugs[7]; this approach is analogous to the method described in Part I of our discussion for beta-adrenergic receptors. Alternatively, the other method involves the use of subtype selective radioligands to label only the subtype for which the radioligand has higher affinity[8].

[³H]Dihydroergocryptine ([³H]DHE) is an antagonist ligand which binds with equal affinity to alpha₁ and alpha₂ receptors. Using computer modelling of competition curves (see Part I) with [³H]DHE it has been found that prazosin is 10,000 fold more potent at alpha₁ than alpha₂ receptors in rabbit uterus, a tissue which contains about 30% alpha₁ and 70% alpha₂ receptors. Conversely, yohimbine, a plant alkaloid, is several hundred fold alpha₂ selective in that tissue. Selective compounds, such as prazosin and yohimbine yield complex biphasic competition curves whereas a non-selective compound, phentolamine, yields a steep monophasic curve even in the presence of both alpha₁ and alpha₂ receptors. With this methodology human platelets have been found to contain exclusively alpha₂ receptors whereas rat liver membranes contain predominantly alpha₁ receptors.

Alternatively, putatively selective radioligands can be used to label exclusively alpha₁ or alpha₂ receptors. Snyder and his collaborators have identified heterogeneous mixtures of alpha receptors in brain membranes. Originally these investigators suggested that this heterogeneity was explained by discrete agonist and antagonist states of the alpha receptors. It was noted that agonists competed with higher affinity than antagonists at sites labelled by tritiated agonists, for example, [³H]clonidine, [³H]epinephrine and [³H]norepinephrine; whereas antagonists competed with higher affinity than agonists at sites labelled with the antagonist [³H]WB4101. More recently, these workers have identified the 'agonist state' with alpha₂ receptors and the 'antagonist state' with alpha₁ receptors. It is not yet clear whether the alpha₁ selectivity of [³H]WB4101 in brain membranes will apply in other tissues to the extent that exclusively alpha₁ receptors and no alpha₂ receptors are labelled by this drug. Indeed, it has recently been reported that [³H]WB4101 binds to the entire alpha receptor population with apparently uniform affinity in rabbit uterus[9]. Therefore, [³H]WB4101 could not be utilized in that tissue to label alpha₁ receptors selectively. Similarly, caution is indicated in utilizing [³H]agonists to label alpha₂ receptors since it may be that these ligands are labelling only a fraction of the alpha₂ receptor population (see discussion below on guanine nucleotide regulation of alpha receptors for details). [³H]Prazosin appears to be a very good ligand for studying alpha₁ receptors. These findings are summarized in Table I.

The complexities involved in the interpretation of radioligand binding studies to alpha receptors are well illustrated by an interesting recent example concerning alpha-adrenergic receptors in liver membranes. Several earlier reports indicated a density of about 1–1.5 pmol mg⁻¹ of alpha-adrenergic receptor binding sites in purified liver plasma membranes as assessed by [³H]DHE binding. Detailed competition curves with prazosin indicate that these sites are about 80% alpha₁ receptors and about 20% alpha₂ receptors. Subsequently, Exton and collaborators[10] found that [³H]epinephrine labelled with high affinity only about 100 fmol mg⁻¹ of sites in such membranes. These high affinity [³H]epinephrine sites were also guanine nucleotide sensitive (see below). In addit-

ion, it was found that considerably more sites were labelled by [³H]epinephrine with lower affinity; however, it was difficult to quantify these exactly. The authors contended that the much smaller number of high affinity [³H]epinephrine sites, as opposed to the [³H]DHE sites, represented 'physiological agonist' alpha receptors whereas the greater number of sites labelled by [³H]DHE represented some 'non-physiological antagonist' form of the receptor[10].

A more plausible explanation of these data is suggested by recent studies. Physiological alpha-adrenergic responses in liver cells, such as glycogenolysis, are mediated by alpha₁ receptors. Prazosin is several orders of magnitude more potent an antagonist of such responses than is yohimbine as it is also in competition for [³H]DHE sites in liver membranes. By contrast, when competition curves of prazosin and yohimbine versus [³H]epinephrine are performed, it is found that at low concentrations of [³H]epinephrine, yohimbine is *more* potent than prazosin indicating that predominantly alpha₂ receptors are being labelled. At higher concentrations of [³H]epinephrine, prazosin becomes pro-

TABLE I. Radioligands used for alpha-adrenergic receptor identification

Radioligand	Receptor specificity	Comments
[³H]Dihydroergocryptine	Equal affinity at alpha₁ and alpha₂ receptors	Antagonist; labels entire alpha-receptor population
[³H]Prazosin	Alpha₁-selective	Antagonist; likely to be generally useful in labelling exclusively alpha₁ receptors
[³H]WB4101	Alpha₁-selective	Antagonist; characterized as alpha₁ selective in brain
[³H]Clonidine [³H]Epinephrine [³H]Norepinephrine	Alpha₂ receptors – high affinity state	Agonists; these agonists likely label predominantly the alpha₂ high affinity state (see text). At higher concentrations they may also label alpha₁ receptors and the lower affinity form of alpha₂ receptors

gressively more potent, but with a shallow displacement curve indicating an interaction with two types of receptors, namely alpha$_1$ and alpha$_2$. Thus, the high affinity [^{3}H]epinephrine sites, labelled at low [^{3}H]epinephrine concentrations appear to be predominantly, if not exclusively, alpha$_2$ receptors. The much greater number of [^{3}H]DHE sites represent alpha$_1$ receptors which are the major physiological subtype in liver membranes. It appears that at higher [^{3}H]epinephrine concentrations a much greater number of alpha$_1$ receptors can also be labelled. However, the affinity of [^{3}H]epinephrine for alpha$_1$ receptors is probably in the range of ~ 100 nM making it experimentally difficult to use [^{3}H]epinephrine to label alpha$_1$ receptors. The importance of the [^{3}H]DHE binding sites in liver membranes is further suggested by the fact that another ligand, [^{3}H]-phenoxybenzamine[11], labels about the same number of sites as [^{3}H]DHE. Thus, in this particular case a seemingly paradoxical situation is observed where under certain experimental conditions the *'physiological'* ligand [^{3}H]epinephrine appears to label sites (alpha$_2$ receptors) which do not mediate the major physiological responses in the liver. The anatomical location and function of the alpha$_2$ receptors labelled by low concentrations of [^{3}H]epinephrine in liver membranes remain to be determined. In analogy with several other systems, however, it may be that these alpha$_2$ sites are linked to adenylate cyclase.

Guanine nucleotide regulation

Guanine nucleotides appear to regulate some alpha receptors in an agonist-specific fashion quite analogous to that previously described for the beta-adrenergic receptors. Alpha$_1$ receptors, thus far, have not been shown to be affected by guanine nucleotides. The alpha$_2$ receptors, which are often coupled in an inhibitory fashion to adenylate cyclase, however, do appear to be regulated by nucleotides. Thus, in a fashion very similar to that for the beta-adrenergic receptors, agonist displacement curves in alpha$_2$ systems are shifted to lower affinity by guanine nucleotides. Antagonist displacement curves are unaffected. Specifically, it can be observed that agonist displacement curves of an antagonist ligand, such as [^{3}H]DHE, are shallow and become steeper and shifted to the right when guanine nucleotides are added. Computer modelling techniques indicate that alpha$_2$ agonists discriminate between high and low affinity states of the alpha$_2$ receptors. In the presence of guanine nucleotides there appears to be an interconversion of the high into the low affinity form of the receptor such that under these conditions only the low affinity form is observed[3]. It can also be shown that guanine nucleotides are required for the alpha-adrenergic-mediated inhibition of adenylate cyclase as, for example, in human platelet[12]. Thus, there appears to be a concerted mechanism whereby guanine nucleotides both destabilize a high affinity alpha-adrenergic receptor complex while also leading to the *inhibition* of adenylate cyclase. This suggests a strong analogy with the mechanisms described above for beta-adrenergic *stimulation* of adenylate cyclase. How such seemingly similar mechanisms can account for stimulation of adenylate cyclase in the one case (beta) and inhibition in the other (alpha) remains an unsolved problem.

Since adrenergic agonists induce a high affinity, guanine nucleotide sensitive form of the alpha$_2$ receptors, such agonists display quite high overall affinity for these receptors in membrane preparations washed free of endogenous GTP. As a result, each of the agonist ligands thus far described, such as [^{3}H]clonidine, [^{3}H]epinephrine and [^{3}H]norepinephrine appear to label predominantly alpha$_2$ receptors. However, under the usual experimental conditions they appear to label predominantly the high affinity state of these receptors. Depending on the tissue and the assay conditions used, this represents a variable fraction of the total alpha$_2$ receptor population. This is usually in the range of 50–70% of the total alpha$_2$ receptor population. Thus, the use of such [^{3}H]agonists to label alpha$_2$ receptors may

TABLE II. Some factors regulating alpha-adrenergic receptor binding sites

Regulatory agent or condition	Effect on receptor
Alpha-adrenergic catecholamines	Decrease the number
Denervation	Increase the number
Hyperthyroidism	Decrease the number, affinity
Estrogen	Increase the number
Thrombocytosis	Decrease the number

be complicated by the fact that only a portion of the alpha$_2$ receptor population will be labelled.

Physiological regulation of alpha-adrenergic receptors

There is considerably less information available about the physiological regulation of alpha-adrenergic receptors compared with beta-adrenergic receptors (see Table II). Several similar mechanisms, however, have been documented. For example, when platelets (which contain alpha$_2$ receptors) are incubated with a catecholamine agonist, such as epinephrine, they become progressively less responsive to the aggregating effect of epinephrine[13]. This 'desensitization' is accompanied by a fall in the number of alpha-adrenergic receptors measured with [^{3}H]DHE. This phenomenon appears to be at least superficially quite similar to the desensitization described earlier for beta-adrenergic receptors. A similar desensitization phenomenon accompanied by a fall in [^{3}H]DHE binding was earlier de-scribed for the alpha-adrenergic receptors coupled to potassium efflux in dispersed rat parotid acinar cells[14]. Conversely, it had been shown that, under some circumstances, denervation, either by surgical or chemical means, may lead to an increase in alpha-adrenergic receptor binding thus providing a potential mechanism for supersensitivity.

Thyroid hormones also appear to regulate the alpha-adrenergic receptors in a fashion which may be opposite in direction to that previously found for beta-adrenergic receptors. Whereas in the rat heart hyperthyroidism increases beta-adrenergic receptor binding, it appears to decrease [^{3}H]DHE binding to alpha-adrenergic receptors. The effects of hypothyroidism on alpha-adrenergic receptor binding are less clear at present, although decreases in alpha receptor binding may also occur in the hypothyroid state.

Other examples of physiological regulation of alpha receptors are the effects of sex steroid hormones on alpha-adrenergic receptor binding. Several groups have reported that estrogens appear to increase alpha-adrenergic receptor number in rabbit uterus whereas progesterone administration decreases the number of alpha receptors. Interestingly, estrogen treatment decreases the number of alpha$_2$ receptors in rabbit platelets. Thus, estrogens seem to cause opposite directional changes in the number of alpha-adrenergic receptors in these two tissues of the same species. The techniques of radioligand binding will undoubtedly prove of great utility in the next few years in attempting to unravel the molecular and cellular mechanisms involved in phenomena such as alpha-adrenergic receptor tachyphylaxis, supersensitivity etc. In such studies, the ability to determine individually the regulatory effects on the different receptor subtypes will be a signif-icant step forward.

Solubilization and purification

Efforts to purify alpha-adrenergic receptors are as yet in their infancy. In the past year two groups have reported successful solubilization of the alpha$_1$-adrenergic receptors of liver membranes. In addition to solubilizing the receptors with digitonin, Wood et al. were able to demonstrate that passage of the soluble extracts over alprenolol Agarose affinity columns led to selective adsorption of beta-adrenergic receptors without adsorption of alpha-adrenergic receptor binding sites[15]. This demonstrated that alpha- and beta-adrenergic receptor binding sites do not appear to reside simultaneously on the same macromolecule. Guellaen et al. pre-labelled the alpha receptors in liver membranes with the irreversible ligand [^{3}H]-phenoxybenzamine and then solubilized the complex. It appeared to have a corrected molecular weight of about 96,000[11]. To date, no work on solubilization of alpha$_2$ adrenergic receptors has been reported. Given the availability of current techniques, the next few years should bring rapid progress in the purification and characterization of both subtypes of alpha-adrenergic receptors.

Clinical studies

As described above, human platelets contain alpha-adrenergic receptor binding sites which appear to be of the alpha$_2$ variety. The ready availability of these human cells containing alpha$_2$ receptors makes them attractive to investigators interested in studying physiological and patho-physiological regulation of alpha receptors. As noted earlier, a major initial concern of all such studies is the relationship, if any, of the properties of the alpha receptors in platelets to those of alpha receptors in other tissues. A great deal of additional data will be necessary to evaluate this point adequately. At present the only reported clinical finding with this system is the observation that in certain forms of thrombocytosis in which the platelets are only poorly aggregated by epinephrine, there is a decreased alpha-adrenergic receptor number.

Conclusions

The availability of direct radioligand binding studies as an approach to investigating alpha- and beta-adrenergic receptors has provided a great impetus to research in adrenergic pharmacology over the past few years. It appears likely that the next few years will yield even greater insights into the molecular and physiological regulation of these receptors and the ways in which they function to modulate and mediate the physiological effects of catecholamines.

Reading list

1 Williams, L. T. and Lefkowitz, R. J. (1976) Science 192, 791–793

2 Greenberg, D. A., U'Prichard, D. C. and Snyder, S. H. (1976) Life Sci. 191, 69–76

3 Hoffman, B. B., Mullikin-Kilpatrick, D. and Lefkowitz, R. J. (1980) J. Biol. Chem. 255, 4645–4652

4 Langer, S. F. (1976) Clin. Sci. Mol. Med. 51, 423S–426S

5 Starke, K. (1977) Rev. Physiol. Biochem. Pharmacol. 77, 1–124

6 Berthelson, S. and Pettinger, W. A. (1977) Life Sci. 21, 595–606

7 Hoffman, B. B., DeLean, A., Wood, C. L., Schocken, D. D. and Lefkowitz, R. J. (1979) Life Sci. 24, 1739–1746

8 U'Prichard, D. C. and Snyder, S. H. (1979) Life Sci. 24, 79–88

9 Hoffman, B. B. and Lefkowitz, R. J. (1980) Biochem. Pharmacol. 29, 1537–1541

10 El-Refai, M. F., Blackmore, P. F. and Exton, J. H. (1979) J. Biol. Chem. 254, 4375–4386

11 Guellaen, G., Aggerbeck, M. and Hanoune, J. (1979) J. Biol. Chem. 254, 10761–10768

12 Jakobs, K. H., Saur, W. and Schultz, G. (1978) FEBS Lett. 85, 167–170

13 Cooper, B., Hardin, R. I., Young, L. H. and Alexander, R. W. (1978) Nature (London) 274, 703–706

14 Strittmatter, W. J., Davis, J. N. and Lefkowitz, R. J. (1977) J. Biol. Chem. 252, 5478–5482

15 Wood, C. L., Caron, M. G. and Lefkowitz, R. J. (1979) Biochem. Biophys. Res. Commun. 88, 1–8

Actions of hormones and neurotransmitters at the plasma membrane: inhibition of adenylate cyclase

Karl H. Jakobs and Günter Schultz

Pharmakologisches Institut der Universität Heidelberg, Im Neuenheimer Feld 366, D-6900 Heidelberg, F.R.G.

The effects of hormones and neurotransmitters which act via membrane-bound receptors involve the generation of intracellular signals (second messengers), which are formed at the membrane and transfer the information to the sensitive regulated cellular system(s). Various hormonal effects have been shown to be mediated by increased formation of cyclic AMP via stimulation of the membrane-bound adenylate cyclase[1]. Hormonal actions that appear to be mediated by this mechanism include for example β-adrenergic effects of catecholamines, effects of various polypeptide hormones such as glucagon and ACTH, effects of some prostaglandins at specific target tissues and various effects of other locally acting hormonal factors such as adenosine and histamine. The coupling of the receptors for these hormones to adenylate cyclase has been extensively studied and is partially understood, although the components of the system and their interactions are far from being elucidated. The system consists of at least three components, i.e., the hormone receptor, the catalytic moiety of adenylate cyclase and a guanine nucleotide binding protein[2,3].

Stimulation of adenylate cyclase involves a GDP–GTP exchange at the binding protein induced by the hormone–receptor interaction as an initiating step and a termination by an inherent GTPase, which is inhibited by cholera toxin-induced ADP-ribosylation of this protein[4].

Through different classes of membrane-bound receptors, the above factors and other hormones and neurotransmitters cause cellular responses that are not mediated by increased cyclic AMP formation. This group includes for example, α-adrenergic effects of catecholamines, (muscarinic) cholinergic effects of acetylcholine, effects of some polypeptide hormones such as vasopressin and angiotensin II, and some effects of prostaglandins, adenosine and related nucleotides and of morphine and natural opiates. With many of these hormonal stimulations, several cellular changes have been observed, which may have the character of signals. An increase in cytoplasmic calcium ion concentration appears to be the most important intracellular signal in many of these hormonal effects[5]. Changes in phosphatidylinositol

TABLE I. Inhibition of adenylate cyclase by hormonal factors: cellular systems and requirements.

Hormonal factor	Cellular system	Requirements*	
		GTP	Na$^+$
α–Adrenergic agonists	Human platelets	+	?
(via α$_2$-adrenoceptors)	Rabbit platelets	+	?
	N × G hybrid cells†	+	+
	Hamster fat cells	+	+
	Human fat cells	+	+
	Rat liver	+	+
	Rat kidney	+	+
	Rat parotid	?	?
Cholinergic agonists	Canine myocardium	+	+
	Rabbit myocardium	+	+
	N × G hybrid cells†	+	+
	Rat parotid	?	?
Opiates	N × G hybrid cells†	+	+
Adenosine	Rat fat cells	+	+
	Hamster fat cells	+	+
ADP	Human platelets	+	?
Prostaglandin E$_1$, E$_2$	Hamster fat cells	+	+
	Rat fat cells	+	+
	Human fat cells	+	+
Dopamine	Pituitary adenoma	?	?
Angiotensin II	Rat liver	+	+
Nicotinic acid‡	Hamster fat cells	+	+
	Rat fat cells	+	+

* +, requirements for GTP or Na$^+$ have been shown; ?, requirements have not been studied or reported.

† N × G hybrid cells, neuroblastoma × glioma hybrid cells.

‡ Although nicotinic is not an established hormonal factor, it has been integrated in this table, since it shows the same characteristics as hormones in the adenylate cyclase studies.

turnover are assumed to be primary to the calcium movements[6], whereas a release of arachidonic acid may be due to the increased cytoplasmic calcium concentration. The increase in cellular cyclic GMP levels found in many of these hormonal effects appears to be related to the increased metabolism of arachidonic acid and to the generation of free radicals during this process[7]. In addition to these changes, a decrease in cellular cyclic AMP levels has often been observed in the course of these hormone actions[1].

Until recently, the enzymatic basis of the hormone-induced decreases in cyclic AMP levels was not clear, i.e. whether it is due to decreased cyclic AMP formation by the adenylate cyclase or to increased degradation of cyclic AMP by stimulation of cyclic nucleotide phosphodiesterase(s). Only insulin has been shown to be able to increase the activity of a phosphodiesterase (the low K_m form) in fat cells and liver[8]. During the last three to four years, many of the other hormonal factors have clearly been demonstrated to reduce cyclic AMP formation by inhibition of the membrane-bound adenylate cyclase (Table I). Receptor-mediated inhibition of adenylate cyclase in membrane preparations has been described for α-adrenergic and muscarinic cholinergic agonists in several tissues, for some prostaglandins, adenosine and also for nicotinic acid in fat cells from different species, for morphine and other opiates in neuronal cells, for ADP in human platelets and for angiotensin II in rat liver (see Refs 9 and 10).

All these inhibitory hormonal effects on adenylate cyclase share some common features:

1. Hormone agonists reduce adenylate cyclase activity without an apparent lag phase, and antagonists (so far available) immediately turn off the inhibitory effects. There is a close correlation between data obtained in binding or physiological response studies and adenylate cyclase inhibition with regard to the potency order of hormonal agonists and antagonists.

2. Inhibition of the adenylate cyclase is seen with the unstimulated and hormone-stimulated forms of the enzyme to about the same extent (40–90%). The enzyme activated by fluoride is less sensitive to an inhibitory attack.

3. In most of the systems studied, GTP, which has been shown to be necessary for hormone-induced adenylate cyclase stimulation[2,3], is also required for hormone-induced inhibition. The GTP concentration for inhibition is usually 5- to 10-fold higher than that necessary for hormonal stimulation. Non- or slowly hydrolysable GTP analogues largely reverse or abolish the hormone-induced inhibition. In contrast, GTP plus cholera toxin, a treatment which is similar to stable GTP analogues in causing apparent persistent activation of the enzyme, does not prevent hormone-induced inhibition. As shown in two systems, i.e. human platelets and rabbit heart, inhibitory hormones largely counteract GTP-induced activation in the presence of cholera toxin.

4. As well as GTP, which is involved in hormone-induced stimulation and inhibition of the enzyme, in most systems sodium or other monovalent cations are required for the demonstration of hormone-induced adenylate cyclase inhibition. As shown in fat cells, liver and heart, sodium ions activate the enzyme in the presence of GTP, and inhibitory hormones appear to counteract this sodium-induced increase in activity. The effective sodium concentration is in the range 20–200 mM.

There are several possibilities for the mechanism of hormone-induced adenylate cyclase inhibition and especially the roles of GTP and sodium ions in this process (Fig. 1). Guanine nucleotides have been shown to modify the binding of hormone agonists to receptors that mediate adenylate cyclase stimulation as well as to other receptors that mediate inhibition[11]. Similar effects of sodium ions have been described only for hormone receptors involved in enzyme inhibition[12]. There is only one exception in which an effect of sodium has also been described on β-adrenoceptors. Up to now, the relationship between the GTP-dependent coupling of stimulatory hormone receptors to adenylate cyclase and that of the GTP-dependent coupling of inhibitory hormone receptors is not clear. Inhibitory hormones and neurotransmitters may inhibit adenylate cyclase by stimulation of the GTPase inherent in the system, which terminates hormone-induced enzyme stimulation. On the other hand, there may be a separate unidentified guanine nucleotide regulatory system which is responsible for hormone-induced inhibition. It is also not clear what the sodium-sensitive step in this hormone-dependent regulation of the enzyme is and whether the sodium-sensitive component

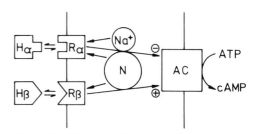

Fig. 1. Regulation of adenylate cyclase activity by stimulatory and inhibitory hormones. H, hormone; R, receptor; α and β, α- and β-adrenergic, representing, respectively, examples of inhibitory and stimulatory hormones; N, regulatory guanine nucleotide binding protein(s); AC, catalytic moiety of the adenylate cyclase.

faces the extracellular or the intracellular space. In the latter case, the system would be coupled by hormone-induced sodium influx, which may increase the sodium concentration locally to that required *in vitro* for inhibition. It may be tentatively speculated that sodium ions somehow interfere with the binding and, thereby, the action of a divalent cation at the plasma membrane. In some intact cellular systems, sodium ions have been shown to be necessary for the hormone-induced fall in cyclic AMP levels[13] and also for the hormone-induced physiological responses.

Whatever the underlying mechanism is[9,10], hormone-induced adenylate cyclase inhibition appears to be a general membrane-related mechanism in the action of hormones and neurotransmitters similar to hormone-induced adenylate cyclase stimulation. Studies on inhibition can serve as a useful approach during *in vitro* studies of early membrane-related events in the actions of those hormones and neurotransmitters whose effects are not mediated by an increased formation of cyclic AMP. The close pharmacological correlation between adenylate cyclase inhibition data and those obtained in studies on binding or physiological responses indicates that a common event in the plasma membrane may precede the enzyme inhibition and the generation of other possible signals leading to the overall cellular response to the hormones.

Further studies on isolated components of the adenylate cyclase system and on the generation of the other possible signals in hormone effects will give further insight into this relationship.

Reading list

1 Robison, G. A., Butcher, R. W. and Sutherland, E. W. (1971) *Cyclic AMP*, Academic Press, New York
2 Ross, E. M. and Gilman, A. G. (1980) *Ann. Rev. Biochem.* 49, 533–564
3 Abramowitz, J., Iyengar, R. and Birnbaumer, L. (1979) *Mol. Cell. Endrocrinol.* 16, 129–146
4 Moss, J. and Vaughan, M. (1979) *Ann. Rev. Biochem.* 48, 581–600
5 Berridge, M. (1975) *Adv. Cyclic Nucleotide Res.* 6, 1–98
6 Jones, L. M. and Michell, R. H. (1978) *Biochem. Soc. Trans.* 6, 673–688
7 Spies, C., Schultz, K.-D. and Schultz, G. (1980) *Naunyn-Schmiedebergs Arch. Pharmakol.* 311, 71–77
8 Loten, E. G., Assimacopoulos-Jeannet, F. D., Exton, J. H. and Park, C. R. (1978) *J. Biol. Chem.* 253, 746–757
9 Jakobs, K. H. (1979) *Mol. Cell. Endocrinol.* 16, 147–156
10 Jakobs, K. H., Aktories, K., Lasch. P., Saur, W. and Schultz, G. (1980) in *Hormones and Cell Regulation* (Dumont, J. and Nunez, J. eds), Vol. 4, North-Holland Publ. Co., Amsterdam, pp. 89–106
11 Williams, L. T. and Lefkowitz, R. J. (1978) *Receptor Binding Studies in Adrenergic Pharmacology,* Raven Press, New York
12 U'Prichard, D. C. and Snyder, S. H. (1978). *J. Supramol. Struct.* 9, 189–206
13 Lichtshtein, D., Boone, G. and Blume, A. J. (1979) *Life Sci.* 25, 985–992

Presynaptic receptors and the control of noradrenaline release

Klaus Starke

Pharmakologisches Institut, Universität Freiburg, Hermann-Herder-Strasse 5, D-7800 Freiburg, F.R.G.

The release of noradrenaline elicited by nerve action potentials is modified by numerous substances, for instance metal cations, local anaesthetics, tetrodotoxin, tetraethylammonium and 4-aminopyridine, scorpion and coral venoms, adrenergic neurone blocking agents, reserpine, and false transmitters. These drugs have taught us much about the mechanism of the release, and many effects are obviously important in toxicology or therapy; however, no effect except for that of some cations such as calcium plays a physiological role. For about 12 years we have known that in addition to foreign compounds several substances that occur naturally in the body are able to modify action potential-evoked release of noradrenaline. At least some of these effects do play a physiological role. The first endogenous modulators to be recognized were angiotensin (mid-1960s), acetylcholine (muscarinic effect, 1968), prostaglandins (1969) and noradrenaline itself (α-adrenergic effect, 1971). Others followed rapidly. These substances modulate the release of noradrenaline via receptors that resemble their receptors on, for example, smooth muscle cells. Although the idea that noradrenergic terminal axons possess receptors was implicit in earlier work, it was first explicitly stated by Lindmar et al.[4] who postulated 'that the peripheral adrenergic nerve fibre contains inhibitory muscarine receptors'. We now call such receptors presynaptic even though noradrenergic axons seldom form morphologically specialized synapses. Fig. 1 shows presently known or proposed presynaptic and soma-dendritic receptors of peripheral and central noradrenergic neurones[3,8,10].

Location of presynaptic receptors

When a drug increases or decreases the release of noradrenaline, its most plausible site of action is the noradrenergic neurone itself. Yet, even in experiments with isolated tissues there are alternatives. For instance, in rat brain cortex slices morphine and endorphins reduce the release of noradrenaline elicited by electrical field stimulation. The opiates may act directly on the cortical noradrenergic terminal axons with their varicosities. However, they may just as well act on neighbouring dendrites, nerve endings or glial cells, which then emit a second signal to the noradrenergic axons. Alternatively, they may act on short interneurones which project to the noradrenergic axon, forming axo–axonic synapses. How can we distinguish between these possibilities? One way is to measure radioligand-receptor binding after noradrenaline-specific lesions. The binding of [³H]naloxone to membranes of rat

<u>Soma–dendritic receptors</u> for:
Acetylcholine (nicotine and
    muscarine receptors)
Catecholamines (α–,β–adrenoceptors)
Serotonin   Histamine
Bradykinin   Angiotensin
GABA

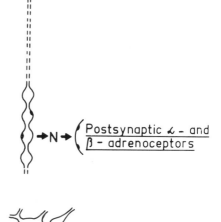

Postganglionic
sympathetic neurone

<u>Presynaptic receptors</u> for:
Acetylcholine (nicotine and
    muscarine receptors)
Catecholamines (α–,β–adrenoceptors,
    dopamine receptors)
Serotonin   Histamine (H_2)
Angiotensin   Endorphins
Prostaglandins
Adenosine
GABA

→N→ (<u>Postsynaptic α – and</u>
      <u>β – adrenoceptors</u>

<u>Soma –dendritic receptors</u> for:
Acetylcholine
Catecholamines
    (α – adrenoceptors)
Endorphins
Substance P
GABA
Serotonin

Central
noradrenergic neurone

<u>Presynaptic receptors</u> for:
Catecholamines
    (α – adrenoceptors)
Endorphins
Prostaglandins
GABA

→N→ (<u>Postsynaptic α–and</u>
      <u>β– adrenoceptors</u>

Fig. 1. Receptor systems of peripheral and central noradrenergic neurones. Soma-dendritic receptors control impulse generation in the perikarya. Presynaptic receptors control the release of noradrenaline (N) per impulse. Release per impulse is increased by activation of presynaptic nicotine receptors, β-adrenoceptors and angiotensin receptors; it is decreased by activation of presynaptic muscarine receptors, α-adrenoceptors, dopamine, serotonin, H_2, morphine, prostaglandin and adenosine receptors; γ-aminobutyric acid (GABA) appears to reduce noradrenaline release from peripheral neurones but to enhance release from central neurones. Activation of presynaptic receptors may also initiate a de novo *release (e.g. nicotine receptors) or modify noradrenaline biosynthesis (angiotensin receptors). Note that a given noradrenergic neurone may possess only some of the receptors summarized here, and that location on the neurone itself is not established for all receptors.*

cerebral cortex is markedly diminished after destruction of the noradrenergic neurones by 6-hydroxydopamine, indicating that in the intact animal these neurones indeed carry specific [³H]naloxone binding sites, i.e. morphine receptors, and that release-inhibiting opiates act there[5]. Analogous findings suggest that in rat hearts muscarine receptors and α-adrenoceptors are constituents of the noradrenergic axons. On the other hand, 6-hydroxydopamine fails to reduce the number of α-receptor ([³H]clonidine) binding sites in rat brain. Negative findings are, however, much less conclusive than positive ones. α-Adrenoceptors on noradrenergic fibres may be too few to contribute significantly to radioligand binding. Moreover, the lesion may lead to a proliferation of postsynaptic α-receptors which balances or even exceeds the loss of presynaptic receptors. Another type of experiment supports location of release-inhibiting α-receptors on noradrenergic varicosities in rat brain. α-Adrenergic agonists diminish the release of noradrenaline even from synaptosomes, i.e. fragments of terminal axons torn out of their normal environment. In this situation, any alternative site of action seems unlikely[7]. In conclusion, for some but not all receptors shown in Fig. 1 there is good evidence that they are presynaptic not only functionally, but also topographically.

Mode of action

The presynaptic receptors summarized in Fig. 1 specifically modify calcium-dependent modes of release, for instance release evoked by electrical stimulation or high potassium concentrations. They do not affect calcium-independent release such as that evoked by tyramine. High potassium depolarizes all varicosities in parallel. Therefore, at least part of the effect of the modulators is probably exerted by regulation of the availability of calcium for electro-secretory coupling in each individual varicosity. There may be additional factors. In contrast to potassium-evoked release, release elicited by electrical stimulation depends on the spread of action potentials along the axons. It seems possible that because of conduction problems the action potential invades only the proximal varicosities, and that presynaptic modulators act by inhibiting or enhancing conduction and, thereby, the recruitment of varicosities[9].

If so, how are the availability of calcium and conduction of the impulse affected? Several biochemical links have been proposed, such as inhibition of presynaptic adenylate cyclase by prostaglandins and activation of the $(Na^+ + K^+)$-ATPase of the neuronal membrane by α-adrenergic agonists. We know nothing for certain. The large number of presynaptic receptors (Fig. 1), when viewed together with their common ability to interfere with electro-secretory coupling, suggests that they share some common sub-receptor structure. Noradrenaline can be released from neurones by reintroduction of calcium after a period of exposure to calcium-free medium (provided a high potassium concentration keeps the neurones depolarized). Muscarinic agonists, α-adrenergic agonists and opiates decrease, and angiotensin II increases, this calcium-induced release, suggesting that they all act by changing the depolarization-induced calcium inward current, and that their receptors are all coupled to potential-sensitive calcium channels as common sub-receptor units[2].

Physiological significance

A question that is raised by even a brief look at Fig. 1 is, why we should have such a multiplicity of presynaptic (and, incidentally, soma-dendritic) receptors. The answer is twofold. Firstly, it seems very likely that some presynaptic receptors have no physiological function whatsoever, simply because they never encounter sufficient concentrations of their endogenous ligands. This may be the case for serotonin receptors for example. Such receptors may be evolutional vestiges which survive because they do us no harm.

Secondly, it seems equally likely that some presynaptic receptors do serve a physiological purpose. Generally speaking, they may be physiological targets of (1) blood-borne agents, (2) substances secreted from neighbouring cells, and (3) the transmitter itself (or other compounds originating from the noradrenergic neurone). We can only speculate as to function (1). Still, presynaptic angiotensin receptors and β-adrenoceptors appear to be sensitive enough to allow significant activation by circulating angiotensin II and adrenaline. It has even been proposed that enhanced noradrenaline release caused by elevated blood levels of angiotensin II or adrenaline might contribute to the pathogenesis of essential hypertension. A well-documented example for possibility (2) is the presynaptic inhibition of postganglionic sympathetic neurones of the heart by adjacent cholinergic neurones. Vagal stimulation *in vitro* as well as *in vivo* diminishes the release of noradrenaline evoked by cardiac sympathetic nerve stimulation. The inhibition is frequency-dependent and abolished by atropine. An analogous interaction seems to occur in other organs where noradrenergic and cholinergic fibres lie side by side. Since, conversely, noradrenaline inhibits the release of acetylcholine (via α-adrenoceptors of the cholinergic neurones), a picture of symmetrical presynaptic inhibition emerges. Noradrenaline and acetylcholine act oppositely on many effector cells. We now know that this classical postsynaptic antagonism is reinforced by mutual presynaptic antagonism between the two divisions of the autonomic nervous system.

Finally – this is function (3) – presynaptic receptors are physiological targets of noradrenaline itself and, hence, links in presynaptic feedback loops. In all tissues studied so far, α-adrenolytic drugs increase the action potential-evoked release of noradrenaline. Their effect reveals that the release is normally braked by an α-adrenergic inhibition that in all likelihood is exerted by released noradrenaline acting on presynaptic α-adrenoceptors. Further arguments support this view. Reserpine depletes noradrenaline stores but does not alter the content of dopamine β-hydroxylase which is located in the vesicles and secreted together with noradrenaline by exocytosis. After treatment with reserpine, little noradrenaline is released, and there should be little α-adrenergic autoinhibition. In fact, under these conditions the release of dopamine β-hydroxylase from the cat spleen is markedly increased. That this is due to removal of autoinhibition is borne out by the fact that phentolamine enhances the release of dopamine β-hydroxylase from spleens of normal, but not of reserpine-pretreated cats[1]. Whereas α-adrenolytic drugs lose their facilitatory effect when there is no feedback, α-receptor agonists lose their inhibitory effect when feedback is maximal, as for instance when the clearance of noradrenaline from the synaptic cleft is depressed by cocaine. At high perineuronal concentrations of endogenous noradrenaline, drugs which are presynaptic partial agonists such as clonidine, and which usually inhibit release, may even cause an increase[6]. Although most of these investigations were carried out *in vitro*, the autoinhibition works *in vivo* as well. For instance, in intact animals α-adrenolytic drugs enhance the tachycardia caused by sympathetic nerve stimulation. The range of conditions under which the feedback mechanism operates *in vivo* deserves further study.

Many noradrenergic neurones possess

presynaptic β-adrenoceptors in addition to α-adrenoceptors. The β-receptors might be expected to mediate a positive feedback mechanism facilitating the release of noradrenaline. However, this feedback cannot be demonstrated regularly, perhaps because the receptors are β_2 and, hence, relatively insensitive to noradrenaline. They may be physiological sites of action of blood-borne adrenaline, as mentioned above.

Some further developments

A corollary of the studies on pre-synaptic α-receptors is the finding that these receptors in many tissues differ from postsynaptic smooth muscle α-receptors. α-Receptors similar to the pre-synaptic ones have subsequently been identified on other structures such as cholinergic neurones, cell bodies of nor-adrenergic neurones, perhaps blood platelets, and recently even certain smooth muscle cells. This type has been called α_2 as opposed to the classical smooth muscle α_1-receptors. The α_2 type is much less sensitive to blockade by prazosin than the α_1 type, but is more sensitive to blockade by yohimbine and its diastereomer rauwolscine. α_1- and α_2-receptors have also been differentiated by radioligand binding. The distinction may be therapeutically significant. For instance, clonidine and α-methylnor-adrenaline, the active metabolite of α-methyldopa, probably produce cardio-vascular depression as well as some side effects (sedation, dry mouth) via α_2-receptors. Prazosin may owe part of its therapeutic properties to selective blockade of α_1-receptors.

The extensive research on presynaptic receptors of noradrenergic neurones has encouraged related studies on other neurones. Cholinergic, dopaminergic, serotonergic, GABAergic and enkephalin neurones all seem to be endowed with presynaptic receptors, and for several receptors a physiological function has been envisaged. Especially interesting are observations indicating that presynaptic autoinhibition may be widespread. More-over, many neurones are inhibited by their own transmitter not only presynap-tically but also in their soma-dendritic region. Self-inhibition by soma-dendritic and presynaptic autoreceptors thus seems to be a rather general neurophysiological phenomenon.

Reading list

1 Dixon, W. R., Mosimann, W. F. and Weiner, N. (1979) *J. Pharmacol. Exp. Ther.* 209, 196–204
2 Göthert, M., Pohl, I. M. and Wehking, E. (1979) *Naunyn-Schmiedeberg's Arch. Phar-macol.* 307, 21–27
3 Langer, S. Z., Starke, K. and Dubocovich, M. L. (eds) (1979) *Presynaptic Receptors,* Pergamon Press, Oxford
4 Lindmar, R., Löffelholz, K. and Muscholl, E. (1968) *Br. J. Pharmacol.* 32, 280–294
5 Llorens, C., Martres, M. P., Baudry, M. and Schwartz, J. C. (1978) *Nature (London)* 274, 603–605
6 Medgett, I. ,C., McCulloch, M. W. and Rand, M. J. (1978) *Naunyn-Schmiedeberg's Arch. Pharmacol.* 304, 215–221
7 Mulder, A. H., de Langen, C. D. J., de Regt, V. and Hogenboom, F. (1978) *Naunyn-Schmiedeberg's Arch. Pharmacol.* 303, 193–196
8 Starke, K. (1977) *Rev. Physiol. Biochem. Pharmacol.* 77, 1–124
9 Stjärne, L. (1978) *Neuroscience* 3, 1147–1155
10 Westfall, T. C. (1977) *Physiol. Rev.* 57, 659–728

New perspectives on the mode of action of antidepressant drugs

Fridolin Sulser

Department of Pharmacology, Vanderbilt University School of Medicine and Tennessee Neuropsychiatric Institute, Nashville, Tennessee 37232, U.S.A.

From acute pharmacological effects to delayed postsynaptic modification of adrenergic receptor function

The classical catecholamine and/or indolealkylamine hypotheses of affective disorders are chiefly derived from studies on the acute pharmacological effects elicited by a number of clinically effective antidepressant drugs and usually do not take into consideration the discrepancy in the time course between biochemical and pharmacological effects elicited by antidepressant drugs within minutes and their clinical therapeutic action which generally requires treatment for several weeks. Indeed, many antidepressant drugs rapidly increase the availability of norepinephrine (NE) and/or serotonin (5HT) either by blocking monoamineoxidase (MAO) or inhibiting the reuptake of the biogenic amines and some of the tricyclic antidepressants are potent antagonists of muscarinic cholinergic receptors, 5HT receptors, dopamine receptors and histamine H_1 receptors. These latter actions are not thought to be responsible for the clinical antidepressant activity as a large number of drugs lacking antidepressant activity are also potent antagonists of these receptors. Moreover, iprindole and mianserin are examples of clinically effective antidepressants with little or no effect on neuronal uptake of NE or 5HT (mianserin actually blocks central 5HT receptors). Obviously, an acute increase in the availability of catechol- and/or indolealkylamines is not directly related to the therapeutic efficacy, and inhibition of reuptake of either NE or 5HT is not an absolute prerequisite for predicting antidepressant activity of non-MAO inhibitor-type antidepressants.

During the last few years, we have been concerned with functional NE receptor interactions in the limbic forebrain and with the pharmacological characterization of the NE receptor-coupled adenylate cyclase system. This adenylate cyclase system shows all the characteristics of a functional central NE receptor system[8]. Studies from our laboratories first demonstrated that chronic but not acute administration of various prototypes of antidepressant drugs and electro-convulsive treatment (ECT) cause subsensitivity of the noradrenergic cyclic AMP generating system in the rat limbic forebrain[17,18]. It was pertinent that the non-uptake inhibitor iprindole shared this effect with both NE uptake and MAO inhibitor-type antidepressants. Table I presents the growing list of antidepressant treatments which have so far been shown to elicit subsensitivity of this noradrenergic receptor system following repeated administration. Since β-adrenergic receptors appear to be a subpopulation of NE receptors, it is not surprising that subsensitivity also develops to the β-adrenergic agonist isoproterenol. Because the selective 5HT uptake inhibitor fluoxetine did not elicit subsensitivity to NE and raphé lesions (unpublished results from this laboratory) did also not alter the sensitivity of this receptor system, it can be concluded that a change in the availability of 5HT *per se* does not change the sensitivity of the cyclic AMP generating system to NE and that the effects of the potent 5HT uptake inhibitor chlorimipramine and also of amitriptyline may be due to the *in vivo* conversion to their secondary amines which are potent inhibitors of the neuronal reuptake of NE. This increased availability of NE could in turn lead to homospecific down-

regulation of noradrenergic receptor function. The observed reduction in the maximum cyclic AMP response without a change in the EC_{50} value (concentration of NE that causes half maximum response) would support the concept of homospecific down-regulation by the agonist NE. However, the subsensitivity developed following chronic administration of iprindole or mianserin is difficult to understand on this basis. Importantly, ECT, one of the most effective treatments for depression with a more rapid onset of action than that of pharmacotherapy, caused rapid subsensitivity to both NE and isoproterenol in limbic forebrain and

TABLE I. Antidepressant treatments which elicit subsensitivity of the NE receptor-coupled adenylate cyclase system in brain

	Subsensitivity to NE and/or Isoproterenol	Reduction in density of β-adrenergic receptors
I. Antidepressant drugs which block uptake of 5HT and/or NE		
Chlorimipramine	+ + +	+ +
Imipramine	+ + +	+ +
Amitriptyline	+ + +	+ +
II. Antidepressant drugs which block predominantly uptake of NE		
Desipramine	+ + +	+ +
Nisoxetine	+ + +	0
III. Drugs which block selectively uptake of 5HT		
Fluoxetine	0	0
Zimelidine	+ + +	?
IV. Antidepressant drugs which do not block uptake of NE and/or 5HT		
Iprindole	+ + +	+ +
Mianserin	+ + +	+ +
V. Antidepressant drugs which block MAO		
Pargyline	+ + +	+ +
Nialamide	+ + +	+ +
Tranylcypromine	+ + +	+ +
VI. Electroconvulsive treatment (ECT)	+ + +	+ +

The data have been obtained either in the author's laboratory or have been computed from the current literature. For further documentation see reference 15.

frontal cortex. Though the potent NE uptake inhibitor nisoxetine has been reported not to induce subsensitivity, recent studies have shown that multiple doses of this short acting antidepressant cause subsensitivity. The results obtained with nisoxetine suggest that homospecific *in vivo* down-regulation of central noradrenergic receptor function depends on persistent exposure of the receptors to high concentrations of NE. The development of subsensitivity to NE is not, however, related to the concentration of the antidepressant drugs in brain tissue, but solely depends on time[18]. It has recently been suggested that acute blockade of histamine(H_2)-sensitive adenylate cyclase by antidepressant drugs might secondarily lead to delayed changes in adrenergic receptor sensitivity[5]. However, since a number of drugs with no antidepressant activity (e.g. promethazine and haloperidol) are potent inhibitors of the enzyme while MAO inhibitor-type antidepressants (pargyline and tranylcypromine) do not inhibit the enzyme, it is clear that this acute effect on histamine-sensitive adenylate cyclase is, unlike the development of subsensitivity to NE, not a common feature of all antidepressants.

Studies utilizing microiontophoretically administered NE have provided electrophysiological evidence for a down-regulation of central noradrenergic receptor activity following chronic but not acute administration of antidepressant drugs. Thus, a statistically significant reduction in the sensitivity of cortical cells to NE occurs following chronic administration of both tricyclics which block the uptake of NE and/or 5HT and following MAO inhibitors[12]. Moreover, while acute treatment with desipramine (DMI) enhanced the inhibitory effect of NE on cerebellar Purkinje firing, chronic treatment with the antidepressant inhibited the effects of NE on the firing rate[14]. Also, a reduction in the *in vivo* accumulation of cyclic AMP following stimulation of the locus coeruleus has been reported in the cerebral cortex in animals chronically treated with antidepressants[6]. Moreover, while a single injection of DMI enhanced AMP response to NE in the rat pineal gland, neither NE nor isoproterenol caused any significant increase in the concentration of cyclic AMP in pineal glands from rats given repeated injections of DMI[11]. Such *in vivo* studies support the view that down-regulation of the NE receptor-coupled adenylate cyclase system by antidepressant drugs is of functional significance. Curiously, long-term treatment of rats with clinically effective tricyclic antidepressants has recently been reported to increase the inhibitory response of forebrain neurons to 5HT applied by microiontophoresis[10]. The mechanism and significance of this effect remain to be elucidated. It is also not yet known whether other types of antidepressant treatment (MAO inhibitors, and ECT) will share this effect with that of tricyclic antidepressants. In any event, in comparing the acute pharmacological actions of antidepressant treatments at presynaptic sites with those on noradrenergic receptor function following treatment on a clinically more relevant time basis, it becomes evident that the delayed therapeutic action is better related in time to the delayed changes occurring at the level of the NE receptor-coupled adenylate cyclase system.

Molecular basis of noradrenergic subsensitivity elicited by antidepressant treatments

Some of the possible mechanisms responsible for drug-induced subsensitivity of the NE receptor-coupled adenylate cyclase system in brain are listed in Table II. Using either ^{3}H-dihydroalprenolol or ^{125}I-hydroxybenzylpindolol to label β-adrenergic receptor sites, it has been demonstrated

TABLE II. Some possible mechanisms underlying subsensitivity of the NE receptor-coupled adenylate cyclase system following antidepressant drugs

1. Change in the affinity of NE to the receptor.
2. Reduction in the density of adrenergic receptors.
3. Reversible uncoupling of the receptor and adenylate cyclase.
4. Changes in the guanine nucleotide regulatory process.
5. Change in the catalytic activity of adenylate cyclase.
6. Change in the activity of phosphodiesterase.

that the subsensitivity of the NE receptor-coupled adenylate cyclase system following chronic treatment with clinically effective antidepressants and with ECT is generally linked to a decrease in the density of β-adrenergic receptors without a change in the binding affinity or K_d value[1,2]. Examples of antidepressant drugs which so far have been shown to reduce the density of β-adrenergic receptors are listed in Table I. Thus, the decrease in the B_{max} value of β-adrenergic receptors appears to be a unique effect caused by antidepressant drugs since it has not been observed following chronic treatment with a wide range of non-antidepressant drugs including pheno-barbital, diphenylhydantoin, diazepam, L-DOPA, haloperidol, promethazine or tripelennamine[13]. Since no apparent alterations in α-adrenergic, serotoninergic or dopaminergic binding (striatum) occur following chronic treatment with anti-depressants or ECT[2,3] the data indicate that antidepressant treatments selectively reduce the number of β-adrenergic receptors. The results are of interest as the regulation of adenylate cyclase activity in brain tissue by β-adrenergic receptors is well documented. Based on studies on agonist activity and on the regulation of the density of β_1 and β_2-adrenergic receptors by the noradrenergic input[7], β-adrenergic receptors in brain involved in neuronal function are primarily of the β_1-subtype. As in other tissues, the fall in the number of functional β-adrenergic receptor sites in brain has been of much lower magnitude than the reduction (subsensitivity) in the NE or isoproterenol stimulated adenylate cyclase activity. This discrepancy may, however, be more apparent than real because catechola-mine-sensitive adenylate cyclase activity is a function of agonist binding whereas the density of β-adrenergic receptors is inferred from antagonist binding to sub-sensitive receptor systems.

Recent results from our laboratory have indicated that subsensitivity of the NE receptor-coupled adenylate cyclase system by antidepressant drugs in brain tissue is not invariably linked to a reduction in the number of β-adrenergic receptors. Dissociations between NE sensitivity of the adenylate cyclase system and the density of β-adrenergic receptors have been observed during development of and recovery from subsensitivity in various brain areas. Clearly, other altera-tions may have occurred such as reversible uncoupling of β-adrenergic receptors and adenylate cyclase. Such changes may precede changes in receptor numbers during the process of development of neuronal subsensitivity. Consequently, it may be inferred that changes in the sensitivity of the NE receptor-coupled adenylate cyclase system to specific agonists provide a more sensitive indicator for functional changes in sensitivity than do changes in the number of β-adrenergic receptors. The elucidation of the probably multiple mechanisms involved in the loss or acquisition of binding sites as well as the mechanisms involved in coupling of the occupied receptor to adenylate cyclase within the phospholipid matrix of neuronal membranes remain exciting areas of future research on the mode of action of antidepressant drugs. Results from Axelrod's laboratory[4] are of great interest in this regard as they have shown that an increase in phospholipid methylation in membranes of reticulocytes through β-adrenergic receptor activation enhances membrane fluidity and coupling of the β-adrenergic receptor with adenylate cyclase. It is tempting to speculate that alterations in microviscosity could, at least in part, be responsible for alterations by antidepressant drugs in neuronal sensitivity of the NE receptor-coupled adenylate cyclase system. Also, the role of potential regulatory agents other than neurohormones in the alteration of noradrenergic receptor function remains to be clarified. As one example, it has recently been found that adrenal corticosteroids can regulate the sensitivity of the NE receptor-coupled adenylate cyclase system in brain[9].

Neurobiological significance of alterations in sensitivity of the NE receptor-coupled adenylate cyclase system in brain elicited by antidepressant drugs

The discovery that most, if not all, antidepressant drugs and ECT cause, upon chronic administration, a down-regulation of central noradrenergic receptor function has provided us with a unique tool to study NE receptor plasticity and adrenergic transmembrane regulation. Both the *in vivo* exploration of the complex central neuronal processes that translate the initial recognition of NE into the final intraneuronal response and the elucidation of the kinetic amplification of the noradrenergic cyclic AMP mediated information-flow will be facilitated. It is becoming clear that an alteration in the sensitivity of the NE receptor-coupled adenylate cyclase system, generally linked to an alteration in the density of β-adrenergic receptors, provides a further dimension of regulatory control of specific biological responses.

The findings that psychotropic drugs or treatments which either precipitate (reserpine) or alleviate (antidepressants and ECT) depressive illness alter the sensitivity of the NE receptor-coupled adenylate cyclase system in cortical and subcortical structures of the brain in opposite directions (supersensitivity or subsensitivity respectively) raises new questions on the psychobiology of depressive illness. One is tempted to advance the hypothesis that depression-prone patients may suffer from an inability to regulate or, more precisely, to down-regulate the central noradrenergic cyclic AMP mediated information-flow (inability to adapt optimally to neuronal input) and that successful treatment with antidepressant drugs and ECT depends on the successful induction of subsensitivity of the NE receptor-coupled adenylate cyclase system. Since chronic stress and antidepressant treatments both produce subsensitivity of the NE receptor-coupled adenylate cyclase system, Stone[15] has proposed that antidepressant therapy is a unique form of adaptation to stress in which the various antidepressant treatments mimic the 'desensitizing' action of stress at central noradrenergic receptors.

Work with antidepressant drugs as both pharmacological tools and therapeutic agents has generated continuous excitement since the early days of their discovery. In the early 'sixties we had the findings that the *in vivo* conversion of tertiary to secondary amines of tricyclic antidepressants leads to more potent 'anti-reserpine' compounds and that this antagonistic action depends on the availability of catecholamines. Then came the important discovery of neuronal amine-reuptake mechanisms and their role in the physiological termination of the action of catecholamines and indolealkylamines. The findings that the conversion of tertiary to secondary amines of tricyclic antidepressants altered the affinity and potency with regard to inhibition of 5HT and NE reuptake furnished more specific tools for the analysis of central neuronal function (see reference 16 for overview). The more recent findings on the action of antidepressant drugs on noradrenergic receptor function and thus on the regulatory control of specific biological responses should contribute to a better understanding of the complex neuronal processes that translate the initial recognition of NE into a final intraneuronal response.

Reading list

1. Banerjee, S. P., Kung, L. S., Riggi, S. J. and Chanda, S. K. (1977) *Nature (London)* 268, 455–456.
2. Bergstrom, D. A. and Kellar, K. J. (1979) *Nature (London)* 278, 464–466.
3. Bergstrom, D. A. and Kellar, K. L. (1979) *J. Pharmacol. Exp. Ther.* 209, 256–261.
4. Hirata, F., Strittmatter, W. J. and Axelrod, J. (1979) *Proc. Natl. Acad. Sci. U.S.A.* 76, 368–372.
5. Kanof, P. J. and Greengard, P. (1978) *Nature (London)* 272, 329–333.
6. Korf, J., Sebens, J.B. and Postema, F. (1978) *Abstracts 4th Intn. Cong. Pharmacol.,* Paris, Abstract 291.
7. Minneman, K. P., Dibner, M. D., Wolfe, B. B. and Molinoff, P. B. (1979) *Science* 204, 866–868.
8. Mobley, P. L., Mishra, R. and Sulser, F. (1979) In: *Catecholamines; Basic and Clinical Frontiers,* E. Usdin, I. J. Kopin and J. Barchas (eds), Pergamon Press, New York, pp, 523–526.
9. Mobley. P. L. and Sulser, F. (1979) *Nature (London)* (in press).
10. Montigny, C. de and Aghajanian, G. K. (1978) *Science* 202, 1303–1306.
11. Moyer, J. A., Greenberg, L. H., Frazer, A., Brunswick, D. J., Mendels, J. and Weiss, B. (1979) *Life Sci.* 24, 2237–2244.
12. Olpe, H. R. and Schellenberg, A. (1979) *Eur. J. Pharmacol.* (in press).
13. Sellinger, M. M., Frazer, A. and Mendels, J. (1979) *Fed. Proc.* 38, 1924.
14. Siggins, G. R. and Schultz, J. (1979) *Proc. Natl. Acad. Sci. U.S.A.* (in press).
15. Stone, E. (1979) *Res. Commun. Psychol. Psychiat. Behav.* 4, 241–255.
16. Sulser, F. (1978) In: *Handbook of Psychopharmacology,* L. L. Iversen, S. D. Iversen and S. H. Snyder (eds), Plenum Press, New York, pp. 157–197.
17. Vetulani, J. and Sulser, F. (1975) *Nature (London)* 257, 495–496.
18. Vetulani, J., Stawarz, R. J., Dingell, J. V. and Sulser, F. (1976) *Naunyn-Schmiedebergs Arch. Pharmacol.* 239, 109–114.

Dopamine receptor: from an *in vivo* concept towards a molecular characterization

Pierre Laduron

Department of Biochemical Pharmacology, Janssen Pharmaceutica, B-2340 Beerse, Belgium.

Dopamine is not only a precursor of noradrenaline but also a neurotransmitter in certain brain regions, namely the striatum and the limbic system. Thus dopamine is synthesized in specific neurones and is released in the synaptic cleft to elicit a physiological effect after binding on a postsynaptic receptor.

The process that led towards the concept of a dopamine receptor began when the first neuroleptic drugs, chlorpromazine and haloperidol were introduced for the treatment of schizophrenia. Neuroleptics were a guide leading to the dopamine receptor just like α-bungarotoxin was for the nicotinic receptor.

In vivo approach

All neuroleptic drugs have the ability to antagonize stereotypy induced by apomorphine and amphetamine in rat, or to prevent apomorphine-induced emesis in dog[1]. In fact these pharmacological tests indirectly suggest that neuroleptic drugs act on the dopaminergic system. Indeed, apomorphine is known to mimic the effects of dopamine, whilst amphetamine, which produces 'psychosis' in man, elicits dopamine release from nerve endings. It is beyond doubt that the exploration of the dopaminergic system has been greatly facilitated by an ideal agonist-antagonist tool, the tandem apomorphine-neuroleptic, that has formed the staple diet of psychopharmacology.

More direct evidence for an interaction of neuroleptics with the dopaminergic system was provided by the observation that dopamine metabolites, dopac and homovanillic acid (HVA) increase in the brain of animals treated with neuroleptics[2]. Such an enhancement in the dopamine turnover, which was confirmed for many drugs in different ways, was interpreted as a consequence of a dopamine receptor blockade. Thereafter, a neuroleptic drug was definitively associated with the dopamine receptor. This led to the statement that the use of a neuroleptic should be a necessary requirement to look more deeply at the molecular level of the dopamine receptor.

In vitro binding

In 1975, the first binding assay *in vitro* for the dopamine receptor was reported independently by two different groups[3,4]. For this, [³H]haloperidol was used as ligand at a nanomolar concentration and the specific binding on striatal membranes was defined as the difference of the radioactivity bound in the presence of (+)-

butaclamol (a potent neuroleptic) and (−)-butaclamol (its inactive enantiomer). That a haloperidol was originally used for labelling dopamine receptors *in vitro* is not surprising since numerous pharmacological and clinical studies have shown the high potency and the selectivity of this drug[1]. This point is illustrated in Fig. 1; when tested in different binding assays (α- and β-adrenergic, muscarinic, serotonin and dopamine) haloperidol as well as pimozide and still more domperidone, exhibit a much higher affinity for dopamine receptors than for the other ones. Therefore, these drugs may be considered as specific dopamine antagonists. It is obvious that chlorpromazine could never become an appropriate ligand for dopamine receptors because its affinity for α-adrenergic and serotonin binding sites is too high (Fig. 1). [3H]flupenthixol and [3H]clozapine are probably also not selective enough. A new ligand needs to be selective and to possess not only a high affinity for dopamine receptors but also a low one for non-specific binding sites. Among those, [3H]spiperone appears to be the ligand of choice owing to its suitable binding characteristics; high affinity for dopamine receptors, low dis-

sociation rate and very low non-specific binding[5]. Such advantages also make this ligand quite appropriate for studies *in vivo*[6] or for binding *in vitro* with solubilized preparations[7]. However, one has to bear in mind that [3H]spiperone is also able to label serotonin receptors[8]. The use of selective displacers and of different brain regions (frontal cortex being mostly serotonergic and striatum dopaminergic) allowed us to differentiate binding sites for both neurotransmitters[8]. As is the case for all binding assays, numerous criteria must be fulfilled to assess the specificity of the binding such as: regional distribution (lack of binding in non-dopaminergic areas) saturability at a nanomolar range, stereospecificity, drug displacement (perhaps the most important and hence the necessity to use a large number of antagonists and agonists belonging to different chemical classes), correlation with pharmacological tests and subcellular characterization. In this regard, our subcellular fractionation studies have clearly shown that the dopamine receptor was not associated with dopaminergic nerve terminals (presynaptic side) but with postsynaptic membrane structures having the characteristics of

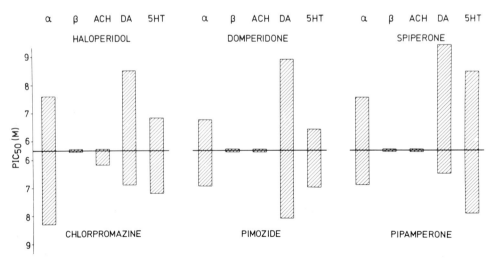

Fig. 1. *Receptor profile of different drugs in various binding assays* in vitro *(α and β: α- and β-adrenergic; ACH: muscarinic; DA: dopaminergic and 5HT: serotonergic). Note that haloperidol, pimozide and domperidone are the most selective dopamine antagonists.*

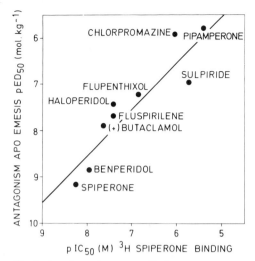

Fig. 2. Correlation between the [³H]spiperone binding in solubilized preparations from dog striatum and the antagonism of the apomorphine-induced emesis in dog.

plasma membranes[9].

As a rule, the interpretation of binding *in vitro* studies needs great caution; we have to remember that sometimes, drugs can displace at non-specific binding sites. Displaceable does not mean specific *per se*. Therefore the choice of an appropriate blank is of the greatest importance. Nevertheless this type of binding assay, like those of many other neurotransmitters, is of great help to pharmacologists, giving a more precise profile for a drug which enables them to rule out the masking effects inherent in pharmacological tests performed *in vivo*.

Towards a molecular characterization

More information about the molecular structure of a receptor can only be provided if binding sites have been previously obtained in a soluble form, which means, in a simpler state only detectable by other techniques than those used for membrane systems. The first attempts to solubilize dopamine receptor from rat striatum were not encouraging[10]; the solubilized material had lost the high affinity properties (except for spiperone itself) normally found on membrane preparations. Occasionally, a

problem arises as work progresses which can only be solved by switching to another animal species. When looking in dog striatum, we were able to solubilize high affinity dopamine receptors using 1% digitonin[7]. The [³H]spiperone macromolecular complex isolated by gel filtration displayed approximately the same affinity towards dopamine agonists or antagonists as the original membrane preparations. The binding was saturable (Kd approximately 4.8 nM), reversible and stereospecific. The macromolecular complex was much more sensitive to thermal inactivation and had a sedimentation coefficient of 9S which corresponds to about 200,000 as the molecular weight although this estimation is probably overestimated due to the formation of micellar structures. Special attention was paid to the criteria of solubilization; it was necessary to ascertain that small pieces of membrane were not taken for solubilized material as has been the case for muscarinic receptors 'solubilized' by high salt concentrations (NaCl 2 M).

Now the question arises as to whether the dopamine receptor in soluble form is really that responsible for the hypermotility, stereotypy and emesis elicited in animals by dopamine agonists and antagonized by neuroleptics or that where neuroleptics act in man to produce antipsychotic effects. There is good reason to believe that it is; indeed, Fig. 2 shows that there is a good correlation between the spiperone binding in solubilized preparations and the antagonism of apomorphine-induced emesis. This strongly suggests that the solubilized dopamine receptor has the same high affinity properties as that involved in the pharmacological and clinical effects of neuroleptics.

More information concerning the molecular structure of the dopamine receptor will be obtained after purification of the receptor. It is to be hoped that such an approach will enable us to provide a

TABLE I.   Comparison of IC$_{50}$-values on adenylate cyclase and haloperidol binding.

Drug	Dopamine-sensitive adenylate cyclase A	[^{3}H]-haloperidol binding B	Ratio A/B
α-flupenthixol	2.3×10^{-8}	$2 \ \times 10^{-8}$	1.2
Chlorpromazine	1.1×10^{-6}	1.3×10^{-7}	8.5
(+)-Butaclamol	$2 \ \times 10^{-7}$	$1 \ \times 10^{-8}$	20
Haloperidol	7.5×10^{-7}	3.6×10^{-9}	208
Pimozide	1.5×10^{-5}	3.2×10^{-9}	4687
Spiperone	2.2×10^{-6}	4.4×10^{-10}	5500
Sulpiride	$> 10^{-3}$	$8 \ \times 10^{-8}$	$> \ 10,000$
Halopemide	$> 10^{-3}$	$1 \ \times 10^{-8}$	$> 100,000$
Domperidone	2.4×10^{-4}	1.4×10^{-9}	142,857

molecular explanation of the fundamental problem presented by the opposite response observed *in vivo* between dopamine agonists and antagonists. It will still be a long time before we can elucidate the molecular events occurring behind the binding site.

Looking again *in vivo*

Another way to ascertain whether a binding site detected *in vitro* is relevant to physiological or pharmacological effects is to characterize them in animals *in vivo*.

One way is to determine the occupancy of the receptor *in vitro* after injection of a drug into the animal. Fig. 3 shows the occupancy of dopamine (striatum) and serotonin receptors (frontal cortex) measured *in vitro* by [^{3}H]spiperone after injection of various doses of pimozide and pipamperone. It is obvious that the former can more specifically bind dopamine receptors in the striatum whereas the serotonin receptors in the frontal cortex were more specifically occupied by pipamperone. This is quite compatible with the *in vitro* binding[8] and pharmacological profiles[1] of these drugs on both receptor systems.

A second way, more sensitive although a little more complicated, is to inject the labelled drug into the animal and then to examine the disposition of labelling in the brain with or without displacement with unlabelled drug[6]. Here again, numerous criteria must be fulfilled to assess the specificity of the binding. Fig. 4 shows the much higher retention of [^{3}H]spiperone in the striatum which was found to be displaceable by various dopamine antagonists or agonists[6]. In contrast to this, the radioactivity in the cerebellum was not only low but was not displaceable. Both procedures may be useful to examine whether or not a drug is able to cross the blood–brain barrier; an example of this has been provided by domperidone, a gastrokinetic and antiemetic drug which, despite a high affinity for dopamine receptors *in vitro* (cf. Fig. 1), is unable to reach brain dopamine receptors *in vivo*. Such binding *in vivo* also seems particularly appropriate for estimating the number of receptors which are really needed for a given agonist to elicit a physiological effect or for a drug to antagonize them.

Is dopamine adenylate cyclase a receptor?

In 1972, a first attempt to identify dopamine receptors biochemically was reported when Greengard's group came to the conclusion that the dopamine sensitive adenylate cyclase represents a possible target for antipsychotic drugs[11]. Indeed, this enzyme occurs specifically in brain dopaminergic areas and only the production of cyclic nucleotide which is stimulated by dopamine was antagonized by certain neuroleptics[12]. Thereafter, the enzyme was considered as a dopamine receptor and even classified as DA_1 subtype[13]. However, when the binding assay for dopamine receptors was reported it became clear that the IC_{50} values of all the antipsychotic drugs for binding correlated better with their pharmacological and clinical potency than the IC_{50} values reported for the enzyme. Table I shows clearly that numerous potent dopamine antagonists are poorly active or not active at all on the cyclase although they compete in the binding assay, sometimes even at a nanomolar concentration. In fact, only the phenothiazines and the thioxanthines were approximately equiactive in both tests. This was not the case of butyrophenones, diphenylbutylamines and benzamides.

Our earlier subcellular fractionation studies provided more direct evidence that the binding site and the dopamine-sensitive adenylate cyclase are two different entities not related to each other[14]. Such a view is at variance with the classical concept according to which the binding site may fuse with the cyclase, both thus representing subunits of the receptor entity.

If one accepts the existence of two different dopamine receptors DA_1 (adenylate cyclase) and DA_2 (binding site), the question arises as to whether both receptors are physiologically or pharmacologically relevant. At the moment, there is no doubt that the antipsychotic drugs act through the binding site and not through the cyclase. Indeed, binding *in vitro* on membrane or solubilized preparations nicely correlates with the pharmacological and clinical potency of neuroleptic drugs; this was not the case for the dopamine-sensitive adenylate cyclase. Moreover, all the pharmacological and behavioural experiments dealing with agonist–antagonist interaction in the brain (hypermotility, turning, stereotypy, vomiting, . . .) can only be explained if such an interaction occurs at the level of the binding site. The binding *in vivo* experiment, especially the data dealing with the subcellular distribution[9], never

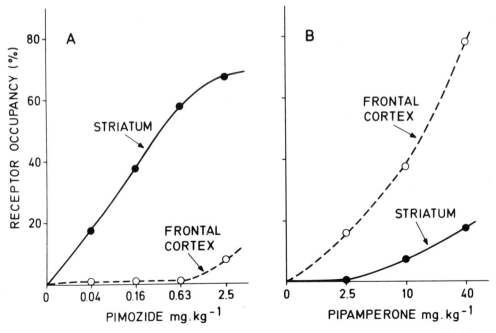

Fig. 3. Binding ex vivo. Rats were injected i.v. with several doses of pimozide and pipamperone and 2 h later, [³H]spiperone binding was performed in different brain regions under in vitro *conditions.*

revealed an association of [³H]spiperone with the cyclase. In fact, up to now, the physiological role of dopamine-sensitive adenylate cyclase is not proven; we believe that even to consider such an enzyme as a receptor remains questionable. If so, one might also consider tyrosine hydroxylase as a receptor; indeed, as with the cyclase, one observes a basal activity which is largely enhanced by pteridine derivatives. Dopamine is thus only a modulator of the adenylate cyclase and in our opinion, the distinction between DA_1 and DA_2 is not justified owing to the lack of experimental data *in vivo* showing a physiological or pharmacological role for the dopamine-sensitive adenylate cyclase. We should consider the binding site as the dopamine receptor and the dopamine-sensitive adenylate cyclase only as an enzyme.

From multiple dopamine receptors to a unitary concept

Like other receptor systems, the dopamine receptor has not resisted the lure of the multiple site concept. When we are faced with anomalous or intriguing results,

the most scientific attitude is first to ask ourselves whether the results are possibly artefactual. In the field of the dopamine receptor, the idea of multiple sites was proposed when dopamine agonists ([³H]dopamine and [³H]apomorphine) were found to label other binding sites than those of antagonists[15]. So, in addition to the DA_1 and DA_2 receptors already discussed, an agonist site, DA_3, was introduced on the basis of rather controversial data (cf. ref. 16). Indeed, although originally [³H]dopamine seemed to recognize the [³H]haloperidol site in rat striatum[3], variations in the binding affinities of drugs for sites labelled by antagonists and agonists were found in calf striatum[4,15]. These observations led to various interpretations such as agonist–antagonist state[4], multiple sites[15], presynaptic localization of the agonist site or identity of the latter with dopamine-sensitive adenylate cyclase (cf. ref. 16). In fact, a systematic re-investigation of the [³H]apomorphine binding demonstrated that it did not differ from that of [³H]spiperone or [³H]haloperidol if the experimental conditions were rigorously controlled to prevent all possible non-specific binding[16].

At the present moment the most plausible hypothesis to describe the dopamine receptor is a unitary concept which does not rule out the possibility of different subunits in the receptor complex itself in order to explain the dualistic effect of agonist and antagonist. The dopamine receptor is unique and the postulated existence of the subtypes DA_1, DA_2, DA_3 . . . or even DA_4, DA_5, DA_6, . . . is probably a short lived fashion.

Conclusion

The dopamine receptor is now well characterized by means of *in vitro* as well as *in vivo* binding assays. Antagonists, [³H]haloperidol and [³H]spiperone seem to be the most appropriate ligands. The macromolecular complex obtained after sol-

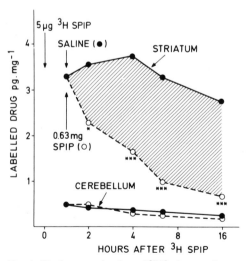

Fig. 4. Displacement in vivo *of [³H]spiperone in rat brain. The hatched part represents [³H]spiperone displaceable by unlabelled spiperone injected 1 h after the labelled ligand.*

ubilization retains the high affinity characteristics of the dopamine receptor identified on membrane preparations.

The dopamine receptor appears to be unique: when it is identified in several different conditions (*in vitro*, membrane and soluble, *ex vivo* and *in vivo*) it presents the same features of the dopamine receptor which has been postulated from pharmacological and behavioural experiments.

Reading list

1 Niemegeers, C. J. E. and Janssen, P. A. J. (1979) *Life Sci.* 24, 2201–2216

2 Carlsson, A. and Lindqvist, M. (1963) *Acta Pharmacol. Toxicol.* 20, 140–144

3 Seeman, P., Chau-Wong, M., Tedesco, J. and Wong, K. (1975) *Proc. Natl. Acad. Sci. U.S.A.* 72, 4376–4380

4 Creese, I., Burt, D. R. and Snyder, S. (1975) *Life Sci.* 17, 993–1002

5 Leysen, J. E., Gommeren, W. and Laduron, P. M. (1978) *Biochem. Pharmacol.* 27, 307–316

6 Laduron, P. M., Janssen, P. F. M. and Leysen, J. E. (1978) *Biochem. Pharmacol.* 27, 317–321

7 Gorissen, H. and Laduron, P. (1979) *Nature (London)* 279, 72–74

8 Leysen, J. E., Niemegeers, C. J. E., Tollenaere, J. P. and Laduron, P. M. (1978) *Nature (London)* 272, 168–171

9 Laduron, P. M., Janssen, P. F. M. and Leysen, J. E. (1978) *Biochem. Pharmacol.* 27, 323–328

10 Gorissen, H. and Laduron, P. M. (1978) *Life Sci.* 23, 575–580

11 Kebabian, J. W., Petzold, G. L. and Greengard, P. (1972) *Proc. Natl. Acad. Sci. U.S.A.* 69, 2145–2149

12 Clement-Cormier, Y. C., Kebabian, J. W., Petzold, G. L. and Greengard, P. (1974) *Proc. Natl. Acad. Sci. U.S.A.* 71, 1113–1117

13 Kebabian, J. W. and Calne, D. B. (1979) *Nature (London)* 277, 93–96

14 Leysen, J. and Laduron, P. (1977) *Life Sci.* 20, 281–288

15 Titeler, M., Weinreich, P., Sinclair, D. and Seeman, P. (1978) *Proc. Natl. Acad. Sci. U.S.A.* 75, 1153–1156

16 Leysen, J. E. (1979) *Commun. Psychopharm.* 3, 387–410

Dopamine receptors and cyclic AMP: a decade of progress

John W. Kebabian and Thomas E. Cote

Experimental Therapeutics Branch, National Institute of Neurological and Communicative Disorders and Stroke, National Institutes of Health, Bethesda, MD 20205, U.S.A.

In the past decade, pharmacologists and biochemists have investigated the biochemical events initiated by stimulation of dopamine receptors. This review summarizes one aspect of these investigations – the regulation by dopamine receptors of adenylate cyclase, the enzyme synthesizing cyclic AMP.

Dopamine-sensitive adenylate cyclase

In 1971, Sutherland and his colleagues, G. A. Robison and R. W. Butcher, noted: 'dopamine may play an especially important role as a transmitter within the CNS, and many of its actions, like those of epinephrine and norepinephrine, may be mediated by changes in the intracellular level of cyclic AMP.' The first experimental evidence indicating that stimulation of a dopamine receptor enhanced cyclic AMP metabolism came from studies of the retina and superior cervical ganglion of the cow; exposure of either bovine tissue to dopamine enhanced the level of cyclic AMP[1,2]. Subsequently, the demonstration of a dopamine-sensitive adenylate cyclase activity in the striatum of the rat brain permitted the characterization *in vitro* of a dopamine receptor from the central nervous system based upon the ability of drugs to mimic or block the stimulatory effect of dopamine upon adenylate cyclase activity of a striatal homogenate[3]. The affinity of a drug for the dopamine receptor could be calculated from mathematical models derived from enzyme kinetics[4].

The physiological process(es) controlled

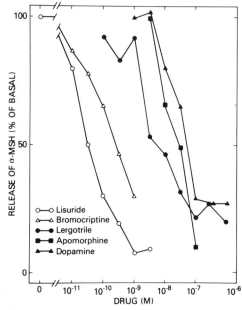

Fig. 1. Dopaminergic agonists inhibit the release of alpha-MSH from dispersed neurointermediate lobe cells.

Incubations were for 180 min in the presence of the indicated concentrations of lisuride (open circles), bromocriptine (open triangles), lergotrile (filled circles), apomorphine (filled squares) or dopamine (filled triangles). Each drug was tested in a separate experiment. In each experiment, the amount of alpha-MSH released was expressed as a percentage of the control release. Modified from Munemura *et al.*, *Endocrinology*, in press[12].

by the dopamine receptor regulating striatal adenylate cyclase is unknown. The electrophysiological and biochemical events regulated by a dopamine receptor in the retina of a fish are currently being characterized by John Dowling and his collegues, Keith Watling and Leslie Iversen[5]. These studies may reveal the type of electrophysiological event controlled by the retinal dopamine-sensitive adenylate cyclase. The insight gained from the retinal experiments may facilitate investigations of the striatum. However, a dopamine-stimulated accumulation of cyclic AMP can initiate physiological events in non-neural tissue. For example, exposure of bovine parathyroid tissue to dopamine enhances the formation of cyclic AMP and stimulates the release of parathyroid hormone; these two responses to dopamine appear to be causally related[6]. At present, the bovine parathyroid provides the best example of a physiological process controlled by a dopamine receptor stimulating adenylate cyclase.

Endogenous guanyl nucleotides regulate adenylate cyclase

The guanyl nucleotide, GTP, confers dopamine-sensitivity to the striatal adenylate-cyclase activity. This role for GTP was first indicated in studies of the subcellular distribution of striatal adenylate cyclase activity. Following centrifugation of the tissue homogenate, the stimulatory effect of dopamine upon adenylate cyclase was lost; the addition of exogenous GTP restored dopamine-sensitivity to the enzyme[7]. Subsequently, the soluble constituent of the striatum removed by repeated centrifugation was identified as GTP[8]. The involvement of GTP in the coupling of a dopamine receptor to adenylate cyclase is consistent with the concept first proposed by Rodbell that following occupancy of a receptor by an agonist, GTP is the intracellular constituent enhancing adenylate cyclase activity.

Multiple dopamine receptors

A dopamine receptor occurs upon the mammotroph of the anterior pituitary gland, the cell synthesizing and secreting prolactin. Some drugs interacting with this pituitary dopamine receptor display 'inappropriate' activity in the assay *in vitro* of dopamine-sensitive adenylate cyclase activity. For example, lergotrile and lisuride are extremely potent dopaminergic agonists when tested upon the mammotrophs of the anterior pituitary gland; however, these drugs are antagonists of the dopamine receptor regulating striatal adenylate cyclase (see 9 for references). The existence of two categories of dopamine receptor may account for the discrepancy between the effects of drugs upon the mammotrophs and the

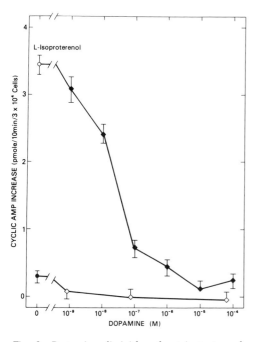

Fig. 2. Dopamine diminishes the L-isoproterenol-stimulated increase in cAMP.

L-isoproterenol (0.1 μM, open circle), dopamine at the indicated concentrations (open diamonds) or dopamine at the indicated concentrations with 0.1 μM L-isoproterenol (filled diamonds) was tested. Modified from Munemura et al., Endocrinology, in press[12].

dopamine-sensitive adenylate cyclase[9]. The nomenclature designating the two types of dopamine receptors as D-1 and D-1 was first proposed by Professor Spano and is now widely accepted (Table I). According to this classification schema the D-1 receptor, the dopamine receptor enhancing adenylate cyclase, is an entity distinct from the D-2 receptor, the dopamine receptor found on the mammotrophs of the anterior pituitary gland[9] and in the intermediate lobe of the pituitary gland (see below). When the subdivision of dopamine receptors was first proposed, the mammotrophs of the anterior pituitary gland provided the best example of a cell which possessed a D-2 receptor. However, direct investigation of the biochemistry of the dopamine receptor on the mammotrophs has not been possible because the heterogeneous composition of the anterior pituitary gland precludes the preparation of a homogeneous population of mammotrophs.

Intermediate lobe

Recent experiments, performed in collaboration with Dr Eskay of NIMH, have investigated the biochemistry and physiology of the dopamine receptor in the inter-

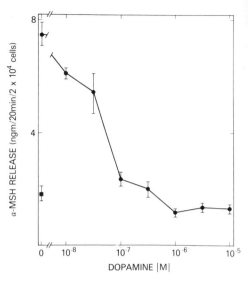

Fig. 3. Dopamine inhibits the L-isoproterenol-enhanced release of alpha-MSH from dispersed cells of the rat neurointermediate lobe. Incubations were described by Munemura et al.[10] in the absence (square) or presence (circles) of L-isoproterenol (1 μM) and the indicated concentrations of dopamine.

mediate lobe of the rat pituitary gland[10, 11]. The intermediate lobe is a homogeneous tissue which synthesizes and secretes alpha-melanocyte stimulating hormone (alpha-MSH) and several other peptide hormones derived from a single precursor molecule, proopiocortin. A beta-

TABLE I. Criteria for the classification of dopamine receptors.

Category of receptor	D-1	D-2
Site of receptor	Bovine parathyroid gland	Rat intermediate lobe
Occupancy by agonist	Enhances formation of cyclic AMP	Does not enhance formation of cyclic AMP; decreases responsiveness of beta-adrenoceptor
Dopamine	Agonist (micromolar potency)	Agonist (nanomolar potency)
Apomorphine	Partial agonist (micromolar potency) Antagonist (nanomolar potency)	Agonist (nanomolar potency)
Dopaminergic ergots	Antagonist (micromolar potency)	Agonist (nanomolar potency)
Selective antagonist	None known at present	(−) Sulpiride

adrenoceptor regulating adenylate cyclase occurs upon the parenchymal cells of the intermediate lobe. Stimulation of this beta-adrenoceptor enhances both the formation of cyclic AMP and the release of alpha-MSH; these two consequences of stimulation of the beta-adrenoceptor appear to be causally related, as is the case for other beta-adrenoceptors regulating a physiological process.

A dopamine receptor also occurs on the parenchymal cells of the intermediate lobe. Dopamine or dopaminergic agonists diminish the spontaneous release of alpha-MSH (Fig. 1). Furthermore, dopamine decreases the consequences of stimulation (i.e. the 'responsiveness') of the beta-adrenoceptor. Thus, dopamine inhibits both the enhanced formation of cyclic AMP (Fig. 2) and the enhanced release of alpha-MSH (Fig. 3) occurring as consequences of stimulation of the beta-adrenoceptor with L-isoproterenol. Either apomorphine or a dopaminergic ergot (e.g. lergotrile) mimics these effects of dopamine; several categories of dopaminergic antagonists, including the substituted benzamides, block the dopamine

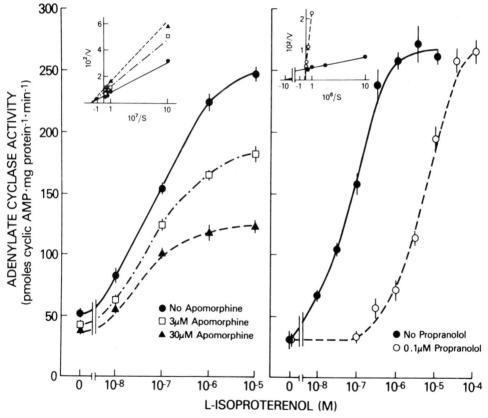

Fig. 4. (Left) Apomorphine non-competitively inhibits the stimulatory effect of L-isoproterenol upon adenylate cyclase activity in a cell-free homogenate of the intermediate lobe of the rat pituitary gland. Incubations were as described by Cote et al[11] in the presence of the indicated concentrations of L-isoproterenol alone (circles) or in combination with 3 μM apomorphine (squares) or 30 μM apomorphine (triangles). Insert: a double reciprocal plot of the L-isoproterenol-stimulated increase in enzyme activity. (Right) Propranolol competitively inhibits the stimulatory effect of L-isoproterenol upon adenylate cyclase activity in a cell-free homogenate of the intermediate lobe of the rat pituitary gland. Incubations were as described by Cote et al.[11] in the absence or presence of 0.1 μM L-propranolol and the indicated concentrations of L-isoproterenol. Insert: a double-reciprocal plot of the L-isoproterenol-stimulated increase in enzyme activity.

receptor[12]. None of the responses to dopamine or the dopaminergic agonists appears to be due to a dopamine-stimulated enhancement of cyclic AMP formation. Therefore, the dopamine receptor in the intermediate lobe has been designated as a D-2 receptor[9].

An element of the initial classification of dopamine receptors was the hypothesis that stimulation of the D-2 receptor did not enhance the formation of cyclic AMP. The presence of a beta-adrenoceptor and a D-2 receptor in the intermediate lobe has permitted the direct experimental investigation of this hypothesis. A beta-adrenergic agonist (e.g. L-isoproterenol) enhances adenylate cyclase activity in a cell-free homogenate of intermediate lobe tissue (Fig. 4, left). In contrast, the dopaminergic agonist, apomorphine does not increase adenylate cyclase activity; however, apomorphine causes a non-competitive decrease in the maximal response of the beta-adrenoceptor to L-isoproterenol. The non-competitive inhibition of the responsiveness of the beta-adrenoceptor caused by apomorphine contrasts with the competitive blockade of the beta-adrenoceptor caused by propranolol (Fig. 4, right[13]).

Some catecholamines stimulate both the beta-adrenoceptor and the D-2 receptor in the intermediate lobe. For example,

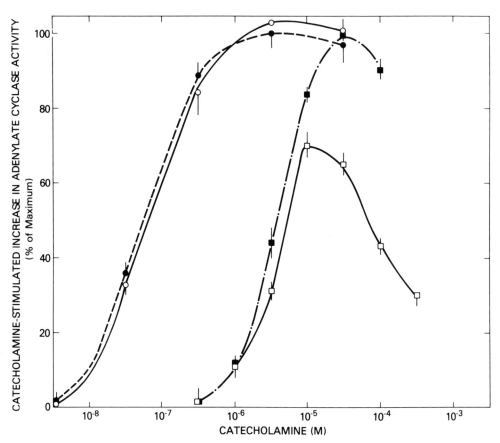

Fig. 5. Stimulation of adenylate cyclase activity by catecholamines. Incubations were as described by Cote et al.[11] in the absence (open symbols) and presence (filled symbols) of 1 μM fluphenazine and the indicated concentrations of L-isoproterenol (circles) or L-norepinephrine (squares). Each catecholamine was tested in a separate experiment; the catecholamine-stimulated increase in enzyme activity observed in each experiment is expressed as a percentage of the increase in enzyme activity caused by 3 μM L-isoproterenol (which was tested in each experiment).

norepinephrine stimulates the beta-adrenoceptor and enhances adenylate cyclase activity in homogenates of the intermediate lobe of the rat pituitary gland[11]; however, the response to norepinephrine is less than the response to isoproterenol (Fig. 5). Fluphenazine, a dopamine antagonist, does not affect the response to L-isoproterenol; however, fluphenazine potentiates the response to norepinephrine. The following 'working hypothesis' accounts for these observations; norepinephrine stimulates both the beta-adrenoceptor and the D-2 receptor; stimulation of the D-2 receptor causes a noncompetitive decrease in the responsiveness of the beta-adrenoceptor resulting in a submaximal enhancement of adenylate cyclase activity. The dopamine antagonist, fluphenazine, blocks the D-2 receptor, reduces the inhibitory constraint upon the responsiveness of the beta-adrenoceptor, and thereby potentiates the beta-adrenergic effect of L-norepinephrine.

Concluding remarks

After ten years of experimental investigations, the hypothesis of Robison, Butcher and Sutherland that the actions of dopamine, 'like those of epinephrine and norepinephrine, may be mediated by changes in the intracellular level of cyclic AMP' continues to provide a useful approach to the investigation of dopamine receptors. Stimulation of the D-1 receptor enhances the formation of cyclic AMP; while in the intermediate lobe of the rat pituitary gland, stimulation of a D-2 receptor decreases the capacity of a cell to synthesize cyclic AMP in response to beta-adrenergic agonists. The occurrence of two categories of receptor for dopamine is reminiscent of the existence of alpha- and beta-adrenoceptors which interact with norepinephrine and epinephrine. In some tissues, stimulation of the alpha-adrenoceptor decreases the capacity of the cells to synthesize cyclic AMP; however, in other tissues, the consequences of alpha-adrenergic stimulation appear to be mediated by calcium. By analogy, stimulation of the D₂ receptor may have biochemical consequences in addition to decreasing the capacity of cells to synthesize cyclic AMP. The insight gained from studying dopamine receptors in simple peripheral tissues may facilitate experimental investigations of dopamine receptors in the brain.

Reading list

1 Brown, J. H. and Makman, M. H. (1972) *Proc. Natl. Acad. Sci. U.S.A.* 69, 539–543
2 Kebabian, J. W. and Greengard, P. (1971) *Science* 174, 1346–1349
3 Kebabian, J. W., Petzold, G. L. and Greengard, P. (1972) *Proc. Natl. Acad. Sci. U.S.A.* 69, 2145–2149
4 Clement-Cormier, Y. C., Kebabian, J. W., Petzold, G. L. and Greengard, P. (1974) *Proc. Natl. Acad. Sci. U.S.A.* 71, 1113–1117
5 Watling, K. J., Dowling, J. E. and Iversen, L. L. (1980) *Neurochemistry* 1, 519–537
6 Brown, E. M., Carroll, R. J. and Aurbach, G. D. (1977) *Proc. Natl. Acad. Sci. U.S.A.* 74, 4210–4213
7 Clement-Cormier, Y. C., Parrish, R. G., Petzold, G. L., Kebabian, J. W. and Greengard, P. (1975) *J. Neurochem.* 25, 143–149
8 Chen, T. C., Cote, T. E. and Kebabian, J. W. (1980) *Brain Res.* 181, 139–149
9 Kebabian, J. W. and Calne, D. B. (1979) *Nature (London)* 277, 93–96
10 Munemura, M., Eskay, R. L. and Kebabian, J. W. (1980) *Endocrinology* 106, 1795–1803
11 Cote, T. E., Munemura, M., Eskay, R. L. and Kebabian, J. W. (1980) *Endocrinology* 107, 108–116
12 Munemura, M., Cote, T. W., Tsuruta, K., Eskay, R. L. and Kebabian, J. W. *Endocrinology* (in press)
13 Cote, T. E., Grewe, C. W. and Kebabian, J. W. *Endocrinology* (in press)

Clinical relevance of dopamine receptor classification

Donald B. Calne

Experimental Therapeutics Branch, National Institute of Neurological and Communicative Disorders and Stroke, Building 10, National Institutes of Health, Bethesda, Maryland 20205, U.S.A.

In developing therapy which involves manipulation of dopaminergic transmission three initial questions must be addressed:

(1) What (if any) dopaminergic mechanism is involved in the pathology of the disease?

(2) What (if any) dopaminergic mechanism is involved in the therapeutic response to treatment?

(3) What (if any) dopaminergic mechanism is involved in the adverse reactions to treatment?

In the following discussion it will emerge that the answers are quite different in various disorders. In formulating a statement on these questions, it will become evident that there are three related issues that require consideration: (1) Why do drugs act (pharmacology)? (2) Where do drugs act (anatomy)? How do drugs act (physiology)?

Drugs which modify dopaminergic transmission are employed in four categories of disease: neurological, psychiatric, endocrinological, and cardiovascular. We shall review these in turn, addressing the questions which have been raised in each case.

Neurological implications

Parkinson's disease. Although there have been several suggestions that Parkinsonism is associated with a widespread disorder of dopaminergic mechanisms, the only irrefutable evidence is limited to involvement of the dopaminergic nigrostriatal system[3]. In this system, there are at least five anatomically distinct sites where dopamine receptors are located[4]: (a) on dopamine cells in the zona compacta of the substantia nigra; (b) on the presynaptic region of dopaminergic axons in the striatum, deriving from the dopamine cells in the nigra; (c) on the presynaptic region of striatal axons coming from neurons situated in the cerebral cortex; (d) on the postsynaptic membrane of striatal neurons; and (e) on the presynaptic regions of axons in the substantia nigra, running from neurons situated in the basal ganglia.

Dopamine receptors at sites (a), (b) and (c) are of the D-2 type, while those at (d) and (e) are of the D-1 type. Since selective agonists for D-2 receptors (bromocriptine, lergotrile and lisuride) are potent therapeutic agents in Parkinsonism, it is probable that efficacy derives, at least in part, from activation of receptors at sites (a), (b) or (c). This conclusion is supported by the fact that dopaminergic blockers which act predominantly at D-2 receptors (such as metoclopramide and molindone) can exacerbate Parkinsonism. The major adverse reactions to dopaminergic therapy in Parkinsonism are nausea, dyskinesia, psychosis and hypotension. Since these effects can also be produced by D-2 agonists, and at least some of them are ameliorated by D-2 blockers, it appears probable that these adverse reactions also stem from D-2 activation. However, with the exception of dyskinesia, the receptors involved in the unwanted effects are unlikely to be located in the nigro-striatal system. The question of how drugs which modify transmission at D-2 receptors work, at a physiological level, is more difficult to answer. Studies involving recording from single neurons indicate that systemic administration of D-2 agonists leads to inhibition of the spontaneous firing of dopamine cells in the substantia nigra, but the complex organization of dopaminergic synapses in the nigrostriatal system makes it difficult to place this observation into any simple hypothesis of extrapyramidal physiology.

Choreoathetosis (dyskinesia). There is no evidence to indicate that brain concentrations of dopamine are abnormal in choreoathetoid dyskinesia. However, drugs which stimulate D-1 and D-2 receptors non-selectively induce choreoathetoid movements in Parkinsonism patients. These drugs are much less prone to produce such movements in subjects who do not have Parkinson's disease, which has led to speculation that the nigrostriatal pathology in Parkinson's disease renders the brain vulnerable to dyskinesia, perhaps by the development of denervation supersensitivity. In accordance with this view, non-selective dopamine blocking drugs, such as phenothiazines and butyrophenones, alleviate choreoathetoid dyskinesia.

The question of whether D-1 or D-2 receptors play the dominant role in generating dyskinesia cannot be answered, though clinical observations suggest that for a given level of anti-Parkinson efficacy, selective D-2 agonists produce less dyskinesia than levodopa[1].

The phenomenon of tardive dyskinesia remains an enigma. This choreoathetoid reaction tends to occur after prolonged therapy with dopaminergic blocking drugs, such as phenothiazines or butyrophenones, and is usually exacerbated when the dose of causal drug is reduced. One explanation proposed to account for these paradoxical findings is that the dopamine antagonists induce dose-dependent reversible blockade of receptors, together with dose-independent, irreversible supersensitivity at those dopamine receptors that are responsible for choreoathetosis.

The tantalizing and important task of identifying the precise site of origin of choreoathetosis has not been achieved, although conventional pathological studies in various choreoathetoid diseases reveal degeneration in the basal ganglia. The physiological basis of choreoathetoid movements is entirely obscure.

Emesis. In discussing the adverse reactions to dopaminergic therapy in Parkinsonism, it was mentioned that D-2 agonists induce nausea. Levodopa is emetic in dogs after gastrointestinal denervation[6], so it seems that the site of action is central, and probably involves direct stimulation of brainstem neurons. Peripheral decarboxylase inhibitors (such as carbidopa), which fail to penetrate the blood–brain barrier antagonize the emetic effect of levodopa. It is, therefore, likely that the relevant receptors are located near the area postrema – a region of the brainstem which is outside the blood–brain barrier. This argument provides a rational basis for the use of dopamine antagonists (such as chlor-promazine) or more specifically D-2 blockers (such as metoclopramide) as antiemetic agents in clinical practice.

Psychiatric implications

Schizophrenia. Attempts to demonstrate a disturbance of dopamine in the brains of schizophrenic patients have failed to establish any definite abnormality. However, it is clear that drugs which stimulate D-2 receptors can produce delusions and hallucinations, while drugs which block D-2 receptors alleviate schizophrenia. From these findings it may be inferred that a net predominance of D-2 activation probably occurs in schizophrenia. The site of the receptors involved has not been identified, and there is no knowledge of the relevant physiology.

Endocrine implications

Hyperprolactinemia. Hyperprolactinemia is being recognized with increasing frequency as a common syndrome, which can be manifested as hypogonadism, infertility and galactorrhoea[8]. It is usually associated with microadenomata of the pituitary in women, or macroadenomata in men. Prolactin is secreted by lactotroph cells of the anterior pituitary, which possess D-2 receptors; activation of these receptors inhibits the release of prolactin. The receptors appear to be normal in adenomata[2], from which it may be inferred that an inadequate quantity of dopamine is gaining access to the pituitary via the hypophyseal portal circulation. The cause of the decrease in dopamine reaching the pituitary is not known.

In the context of considering dopaminergic mechanisms, it should be mentioned that dopamine blocking drugs, such as phenothiazines and butyrophenones, induce hyperprolactinemia.

Hyperprolactinemic syndromes respond extremely well to treatment with selective D-2 agonists; bromocriptine is the most extensively used therapeutic agent. Bromocriptine is so effective in decreasing the plasma concentration of prolactin that it is widely employed to terminate normal lactation.

Acromegaly. Acromegaly is due to excessive secretion of growth hormone by the pituitary, but it is not known whether the primary cause is the pituitary adenoma, or an abnormality of the hypothalamic mechanisms that control the anterior pituitary. The control of the secretion of growth hormone is complex, but normally administration of dopamine leads to an increase in plasma concentrations of this hormone. In acromegaly, however, dopaminergic agents decrease plasma growth hormone levels, and this effect can be achieved with selective D-2 agonists such as bromocriptine[7]. It has been suggested that this reversal of dopaminergic action in acromegaly may derive from dedifferentiation of somato-mammotropic cells to a more primitive lactotroph form. From a practical viewpoint, the important point is that bromocriptine represents the first practical and effective treatment for both the biochemical and clinical features of acromegaly.

Pituitary tumors. One of the most interesting recent observations in the management of pituitary tumors is that adenomata may decrease in size following the administration of bromocriptine. Prolonged treatment with bromocriptine has been reported to induce regression in a proportion of tumors in hyperprolactinemic or acromegalic patients. The mechanism of this action is not understood, but since no other medical treatment is capable of inhibiting the growth of pituitary tumors, we may be on the threshold of important therapeutic developments in an entirely unexpected area.

Cardiovascular implications

Shock syndromes. Shock syndromes are characterized by arterial hypotension sufficient to imperil the vascular perfusion of organs. Although such syndromes are not caused by pathology involving dopaminergic synapses, intravenous administration of dopamine is one of the standard aproaches to therapy. Dopamine exerts its beneficial effects by increasing the force of contraction of the heart, through activation of noradrenaline receptors. Noradrenaline infusion will have a similar action on the heart, but will also produce widespread vasoconstriction which can seriously compromise the renal circulation. In contrast, dopamine dilates the renal vasculature, by stimulating specific receptors whose pharmacological properties imply categorization as D-1 type[5]. Thus dopamine increases the concentration of cyclic AMP in the perfusate of the dog's kidney, and selective D-2 agonists fail to induce renal vasodilatation.

Conclusions

From this review it is apparent that while the D-1 receptor is better understood in the laboratory, the D-2 receptor receives more attention in the clinic. However, it is appropriate to stress that the failure, up to now, to attribute many clinically relevant functions to the D-1 receptor does not in any way imply that this category of receptor is unimportant. More work is needed to elucidate, for example, the role of the high concentration of D-1 receptors in the retina of most mammalian species. To deny a potentially significant function for the D-1 receptor is analogous to disclaiming any role for the lymphocyte prior to modern immunology. D-1 receptors exist, and we need to find out more concerning what they do. One major handicap in this task is the current lack of adequately selective D-1 agonists or antagonists suitable for studies in man.

This survey outlines the breadth of clinical significance deriving from research into dopamine synapses in general and dopamine receptors in particular. Such common problems as emesis or termination of normal lactation are likely to affect the majority of any population, and drugs which act on dopamine receptors have a part to play in their management. In addition, serious clinical disorders such as shock, Parkinsonism, schizophrenia, hyperprolactinemia and acromegaly can undergo significant amelioration in response to manipulation of dopaminergic receptors. From this wide array of applications, it is evident that a multidisciplinary approach to research is desirable, in which therapy is related to synaptic receptors rather than aetiology; treatment is assigned on the basis of pharmacology rather than pathology.

Reading list

1 Calne, D. B., Williams, A. C., Neophytides, A., Plotkin, C., Nutt, J. G. and Teychenne, P. F. (1978) *Lancet* i, 735–738

2 Cronin, M. J., Cheung, C. Y., Wilson, C. B., Monroe, S. E., Jaffe, R. B. and Weiner, R. I. (1980) *J. Clin. Endocrinol. Metab.* 50, 387–391

3 Hornykiewicz, O. (1973) *Fed. Proc., Fed. Am. Soc. Exp. Biol.* 32, No. 2, 183–190

4 Kebabian, J. W. and Calne, D. B. (1979) *Nature (London)* 277, 93–96

5 Nakajima, T., Naitoh, F. and Kuruma, I. (1977) *Eur. J. Pharmacol.* 15, 195–197

6 Peng, M. T. (1963) *J. Pharmacol. Exp. Ther.* 139, 345–349

7 Thorner, M. O., Chait, A., Aitken, M., Benker, G., Bloom, S. R., Mortimer, C. H., Sanders, P., Mason, A. S. and Besser, G. M. (1975) *Br. Med. J.* 1, 299–303

8 Thorner, M. O. (1977) *Clin. Endocrinol. Metab.* 6, 201–222

Receptors and calcium signalling

Michael J. Berridge

Unit of Invertebrate Chemistry and Physiology, Department of Zoology, University of Cambridge, Downing Street, Cambridge CB2 3EJ, U.K.

Most hormones and neurotransmitters exert their effects without entering the cell. They interact with specific receptors which are coupled to various effector or amplifier systems responsible for generating internal signals or second messengers (Fig. 1). The cyclic nucleotides (cyclic AMP and cyclic GMP) and calcium constitute a second messenger triumvirate which presides over the control of many cellular processes. Calcium is particularly important since an increase in the intracellular level of this bivalent cation controls contraction, exocytosis, membrane permeability, metabolism, cell division and a host of other processes. Considering the importance of calcium as a ubiquitous second messenger, we are surprisingly ignorant concerning the transducing mechanisms responsible for generating this signal. This article is mainly concerned with those receptors that are known to act through calcium.

Receptor classification

Before dealing with calcium signalling, I shall briefly summarize some of the main features of the signal processing mechanisms responsible for generating the other second messengers.

In Fig. 1, a representative collection of agonists has been classified with respect to which of the second messengers they employ to exert their effects. One major class operates mainly through cyclic AMP, whereas another major class seems to use calcium and cyclic GMP. There are other receptors such as those for the opiates, where the nature of the second messenger has not been resolved. Such receptors may act indirectly by modulating the ability of other receptors to generate second messengers as will be described later. The nicotinic receptor is a special case in that it does not directly lead to the formation of a second messenger because it is coupled to a sodium channel whose opening leads to a rapid change in membrane potential. However, such voltage changes may result in a calcium signal if the membrane also contains voltage-dependent calcium channels as occurs in the adrenal medulla.

Apart from special cases of this kind, most receptors seem to act directly either through cyclic AMP or through the combined action of calcium and cyclic GMP. As is evident from Fig. 1, certain agonists can operate through either second messenger pathway by acting on separate receptors.

The terminology used to describe these receptors is somewhat confusing. When it became apparent that there was more than one receptor for agonists such as norepinephrine, the Greek alphabet was used to discriminate between them and we have all become accustomed to using α and β to describe the two major adrenergic receptors. When subpopulations of these two major receptor classes were identified, they were distinguished by using numerical subscripts hence we have α_1 and α_2 and β_1 and β_2. Although this represents a fairly rational system, it has not been applied to other receptors where the numerical system has been used to distinguish between the major types of dopamine, histamine and 5-hydroxytryptamine receptors (Table I). Even more confusion arises when one considers the second messenger events associated with each of these receptors. One obvious inconsistency concerns the receptors linked to cyclic AMP which are D_1 for dopamine and H_2 for histamine (Table I). It is to be hoped that as each new receptor is identified, special care will be taken to try and develop a consistent nomenclature which takes into account the second messenger pathway used by that receptor.

Are there common transducing mechanisms?

Each second messenger is generated by a specific effector system (Fig. 1). The effector systems for the cyclic nucleotides are guanylate and adenylate cyclase, whereas that for calcium is represented by a calcium channel or gate located either in the plasma membrane or in a membrane of some internal pool such as the sarcoplasmic reticulum of muscle. One of the intriguing problems of modern pharmacology is to explain how all these different receptors are coupled to this limited range of effector systems. Another important question to consider is whether each receptor is linked to its effector system through a private line or whether all the receptors in each group operate through a common intermediary. Early studies on fat cells, which have several receptors capable of generating cyclic AMP, argued against the existence of private lines of communication because there was no additivity when maximal concentrations of different agonists were presented in combination. It is now apparent that all receptors which operate through cyclic AMP share a common transducing and effector mechanism (Fig. 1). The main feature of this signal pathway is that the specific receptors, which face the outside, are coupled to adenylate cyclase by a GTP-binding protein (N_s), part of which faces the cytoplasm to gain access to GTP[1]. The activation of adenylate cyclase by N_s may not be the sole signal processing pathway for these cyclic AMP-dependent agonists because there is evidence that some of these agonists may increase phospholipid methylation[2]. For example, in reticulocyte membranes the activation of β-receptors stimulates methyl transferases which convert phosphatidylethanolamine into phosphatidylcholine. It has been proposed that this phospholipid interconversion alters membrane fluidity which could then enhance the ability of receptors to

TABLE I. The association of separate receptors with different second messenger pathways.

Agonist	Receptor-mediated effects		
	Increased cAMP	Decreased cAMP	Increased Ca^{2+} and/or cGMP
Acetylcholine	–	Muscarinic	Muscarinic
Norepinephrine	β	α_2 [a]	α_1
Dopamine	D_1	D_2	?
5-Hydroxytryptamine	$5-HT_2$	?	$5-HT_1$
Histamine	H_2	?	H_1

[a] – taken from Ref. 14.

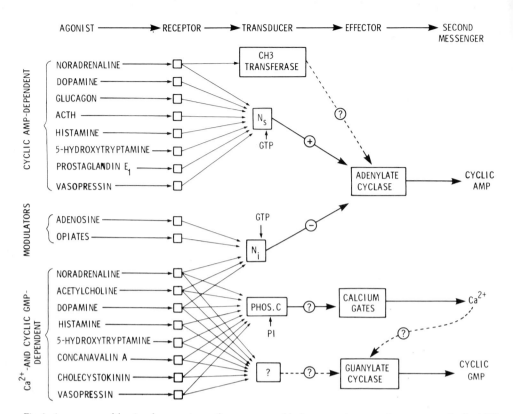

Fig. 1. A summary of the signal processing pathways responsible for generating second messengers. Cyclic AMP-dependent agonists use a GTP-regulatory protein (Ns) to pass information from the receptor to adenylate cyclase. Agonists which act through calcium and cyclic GMP may have access to several transducing systems including an inhibitory GTP-regulatory protein (Ni), phospholipase C (Phos. C) which hydrolyses phosphatidylinositol (PI), and an unknown mechanism which activates guanylate cyclase (see text for further details).

couple to adenylate cyclase[2]. It remains to be seen, however, whether this methylation reaction is a general phenomenon associated with all receptors responsible for generating cyclic AMP.

If we now turn our attention to receptors which act through calcium it is important to consider whether they share a common pathway or whether they too may be linked to separate transducers. It is difficult to provide an unequivocal answer to such a question because we still know so little about how these receptors raise the intracellular level of calcium or activate guanylate cyclase. However, enough evidence has accumulated to suggest that these receptors may operate through different transducing mechanisms (Fig. 1).

The role of GTP-binding protein (Ni)

One transducer associated with these calcium-dependent receptors may have nothing to do with calcium but may exert an inhibitory effect on adenylate cyclase. There are two major types of GTP-regulatory proteins which control the activity of adenylate cyclase. One causes stimulation (Ns) as already described for the cyclic AMP-dependent signal processing system. The second type (Ni) is linked to other receptor classes and leads to an inhibition of adenylate cyclase[1]. Agents which have been found to act through Ni include adenosine, opiates, acetylcholine, α-adrenergic agents and dopamine (Fig. 1). It remains to be seen, however, whether this inhibitory effect on adenylate cyclase represents the sole mode of action of these agonists. To answer this question it is necessary to examine how these various agonists act on their target cells. In adipose tissue for example, adenosine is generated by fat cells during hormonal stimulation and may function as an endogenous modulator whose primary action is to reduce the activity of adenylate cyclase. The opiates may have a similar mode of action in that they modulate the responsiveness of nerve cells to other neurotransmitters by tuning down the activity of adenylate cyclase as has been described in neuroblastoma/glioma hybrid cells[3]. However, in the case of the mammalian salivary gland, an inhibition of adenylate cyclase cannot represent the sole mode of action of acetylcholine and noradrenaline since the latter act mainly through calcium. Based on what we know about the second messenger feedback interactions in the salivary gland[4], a decrease in the level of cyclic AMP will act to lower rather than increase the level of calcium. Such a decrease in cyclic AMP production certainly represents one output from these receptors[5] but is clearly unlinked to the major output signal which is an increase in the intracellular level of calcium[4]. It is difficult to escape the conclusion that these receptors must be linked to alternative effectors such as guanylate cyclase and calcium gates.

One possibility, which has not been fully explored, is that GTP-regulatory proteins, perhaps even N_i, may function as a transducer for more than one effector system. For example, N_i may be linked to a calcium gate in addition to exerting its inhibitory effect on adenylate cyclase. Rodbell[1] has also referred to certain agonists which seem to be associated with GTP-regulatory proteins (N_x) for which effector systems have not been identified.

Cyclic GMP

Most agonists which are thought to act through calcium are also capable of stimulating the formation of cyclic GMP. Guanylate cyclase thus represents an alternative effector system but not much is known about the transducing system which functions in this signal pathway. One possibility is that all these receptors act through some common transducer which then activates guanylate cyclase (Fig. 1). Another possibility is that the activation of guanylate cyclase occurs indirectly through calcium. There are numerous examples to show that the formation of cyclic GMP is dependent on calcium. The precise signal pathway responsible for generating cyclic GMP remains very much a mystery.

Phospholipid metabolism

The signal processing which occurs when external signals are transduced into second messengers occurs within the phospholipid bilayer. In the coupling mechanism for cyclic AMP formation described earlier, processing is carried out by protein–pro-

TABLE II. A summary of some of the stimuli which are known to induce a PI response in specific tissues.

Stimuli	Tissues
Cholinergic (muscarinic)	Avian salt gland
	Pancreas
	Parotid gland
	Smooth muscle
	Lacrimal gland
Adrenergic (α)	Parotid gland
	Smooth muscle
	Hepatocytes
Histamine (H_1)	Smooth muscle
Pancreozymin	Pancreas
Substance P	Parotid
Vasopressin	Hepatocytes
Angiotensin	Hepatocytes
5-Hydroxytryptamine (5-HT$_1$)	Smooth muscle
	Blowfly salivary gland
Concanavalin A	Mast cells
Thrombin	Blood platelets

tein interactions as information is transferred from the receptor to either N_s or N_i and
then to adenylate cyclase. There appears to
be little specific role for phospholipids
other than providing a hydrophobic milieu
in which these proteins can interact. Such a
passive role for the phospholipid bilayer
may be too simplistic a view because there
is growing evidence that agonists can
induce specific phospholipid transforma-
tions which could be important in information transduction. The ability of

β-adrenergic agents to increase phospholipid methylation possibly leading to
changes in membrane microviscosity[2] was
referred to earlier. The class of signal
molecules which act through calcium or
cyclic GMP also exert a profound and
specific effect on phospholipid metabolism
which may play an essential role in information transduction.

The first inkling that certain external
agonists may act by hydrolysing phospholipids emerged from studies on the ac-

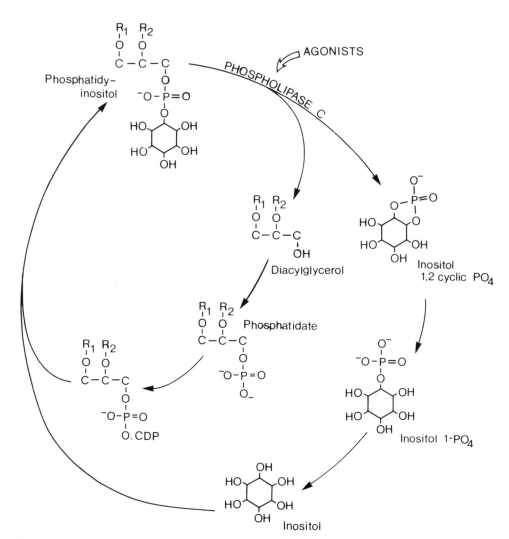

Fig. 2. The metabolic pathways responsible for phosphatidylinositol (PI) metabolism. Agonists increase the turnover rate by stimulating the hydrolysis of PI by phospholipase C.

tion of acetylcholine on pancreas by Hokin and Hokin[6]. Subsequent studies showed that acetylcholine singled out phosphatidylinositol (PI) leaving the other phospholipids unchanged. This specific hydrolysis of PI has been referred to as the PI response and has been observed in many different cells in response to a wide range of stimuli (Table II). Phosphatidylinositol is an acidic phospholipid which is usually a minor membrane component mainly confined to the inner leaflet of the plasma membrane. It is cleaved by phospholipase C to form 1,2-diacylglycerol and inositol 1,2-cyclic phosphate (Fig. 2). These two products are recycled during the resynthesis of phosphatidylinositol (PI). The 1,2-diacylglycerol is rapidly phosphorylated to phosphatidate which then combines with CTP to form the diacylglycerol CDP which finally reacts with inositol to reform PI (Fig. 2). There is a phosphodiesterase and a phosphatase which are responsible for converting inositol 1,2-cyclic phosphate to inositol 1-phosphate and then to inositol.

Evidence for considering that this PI response is an early aspect of receptor function has been advanced by Michell et al.[7-9]. They point to the close correlation that exists between receptor occupancy and the PI response. There is no such correlation when cellular responses are considered because these are usually maximal at very low agonist concentrations as is consistent with the spare receptor hypothesis. The occupation of a small proportion of the available receptors results in a small PI response but a maximal change in cellular activity such as contraction or secretion. Further increases in agonist concentration result in more receptors being occupied with correspondingly larger PI responses but there is no further change in cellular activity. The fact that the breakdown of PI is not related to cellular activity but to receptor occupancy provides further evidence that it is a receptor mediated event.

If the hydrolysis of PI by phospholipase C is a crucial component of the signal pathway then one might expect that the inclusion of such an enzymatic step might slow down the rate at which these signals act. Long latencies are in fact a characteristic feature of receptors that appear to function through the PI mechanism. For example, responses mediated by muscarinic receptors occur very much more slowly than those taking place through nicotinic receptors. There now seems to be strong circumstantial evidence that this hydrolysis of PI is an early event of receptor activation which could thus represent one of the transducing mechanisms linked to such receptors (Fig. 1). What remains in doubt concerns the functional significance of this cleavage of membrane phospholipid. If this PI response is an important general transducing mechanism, it remains to identify the effector system that responds to this acceleration in PI metabolism.

PI metabolism and calcium gating

Michell and his colleagues[7-9] have put forward the interesting hypothesis that this PI response is linked in some way with the generation of a calcium signal. The idea behind this hypothesis developed from the realization that all receptors which gave a PI response seemed also to act through calcium. In examining this hypothesis it is clear that we have to distinguish between two alternative functions (Fig. 3). As outlined earlier, there is little doubt that PI hydrolysis is an important receptor-mediated response but whether this response occurs in parallel with an increase in calcium gating (Fig. 3a) or whether it plays a direct role in generating this signal (Fig. 3b) as proposed by Michell et al. is more difficult to decide. One approach to this problem has been to assess the calcium-sensitivity of the PI response. Michell[8] argued that if the hydrolysis of PI preceded the generation of the calcium signal it should be independent of any increase in the intracellular level of this ion. This prediction has been confirmed in a number of different cells. However, while the insensitivity of the PI response to calcium is a necessary prerequisite for it to function as a transducer for calcium signalling, it does not help to decide between the two alternatives summarized in Fig. 3. One of the major problems encountered in studies on this signal pathway concerns the difficulty of trying to measure calcium permeability. In order to provide evidence for an involvement of the PI response in calcium gating it is imperative to develop techniques for monitoring the rate at which calcium moves across the membrane.

The insect salivary gland

In a recent review on cell surface receptors Michell[8] concluded that 'we should expect that the next key pieces of this jigsaw will come from rather unexpected sources'. One of these unexpected sources may turn out to be the salivary gland of a fly which has an unusual feature of calcium metabolism which can be exploited to provide a simple technique for monitoring membrane permeability to calcium continuously in intact cells[10]. Techniques for continuously monitoring the hydrolysis of PI have also been developed. Using these two techniques, it was shown that 5-hydroxytryptamine (5-HT) stimulates both the hydrolysis of PI and the entry of external calcium. These two processes were always closely related under a variety of conditions. The fact that a PI response is always associated with an increase in calcium entry certainly favours the sequential pathway outlined in Fig. 3b.

In order to assemble more direct evidence for an involvement of PI hydrolysis in calcium gating an attempt was made to remove or lower the membrane level of PI[11]. It was argued that if PI plays an important role then removal of this precursor should seriously curtail calcium gating. Under conditions which cause a progressive depletion of membrane PI, there was a parallel fall in calcium entry[11]. This inactivation of calcium entry was completely reversed when glands were allowed to resynthesize PI. The fact that the ability of 5-HT to regulate calcium entry could be altered simply by varying the level of PI in the membrane does support the hypothesis that hydrolysis of this phospholipid is an important transduction step in the signal pathway leading to the opening of calcium gates (Figs 1 and 3b).

Remaining problems

Although evidence that PI hydrolysis plays a direct role in calcium signalling is beginning to accumulate there are still large gaps in the jigsaw which require filling.

The problem of internal calcium. Certain cells not only use external calcium but are capable of releasing calcium from internal reservoirs. In both liver and pancreas, for example, external receptors can mobilize

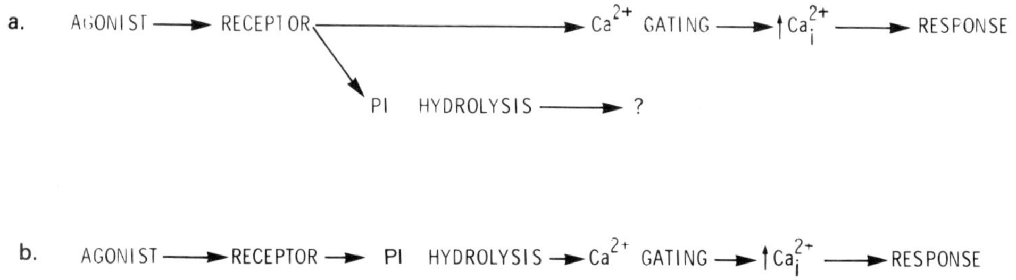

a.   AGONIST $\longrightarrow$ RECEPTOR $\longrightarrow$ Ca^{2+} GATING $\longrightarrow$ $\uparrow Ca_i^{2+}$ $\longrightarrow$ RESPONSE

PI HYDROLYSIS $\longrightarrow$?

b.   AGONIST $\longrightarrow$ RECEPTOR $\longrightarrow$ PI HYDROLYSIS $\longrightarrow$ Ca^{2+} GATING $\longrightarrow$ $\uparrow Ca_i^{2+}$ $\longrightarrow$ RESPONSE

Fig. 3. Two alternative views of the relationship of PI metabolism to calcium gating. In (a), there is no direct relationship and the PI response is a parallel event of unknown function, whereas in (b) it is an essential part of the signal pathway responsible for calcium signalling.

internal calcium in addition to their effects on cell surface calcium gates[15]. While it is reasonable to see how PI hydrolysis in the surface membrane can lead to an entry of external calcium, it is more difficult to envisage how a PI response at the surface could release calcium from internal reservoirs. The way in which information is transferred from the cell surface to internal membrane systems is still unknown. The possibility that inositol 1,2-cyclic phosphate formed during the PI response might perform such a second messenger function has not been properly excluded. We clearly require more information about the nature of internal calcium reservoirs before pro-

gress can be made in elucidating the release mechanism.

The source of PI. If the hydrolysis of PI does turn out to play a crucial role in signal processing, then the source of this phospholipid will become a matter of considerable interest. Since the source of PI used for transduction is the membrane itself, we must assume that precursor PI reaches the receptor sites by lateral diffusion within the membrane. Since phospholipids diffuse rapidly within the plane of the membrane these receptor sites probably have a continuous supply of PI especially when receptors are evenly distributed over the surface of the cell. However, if receptors are

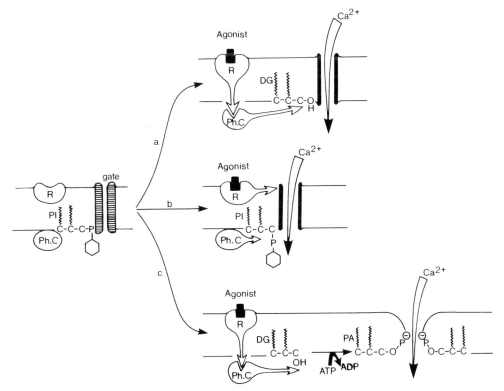

Fig. 4. Three possible models for linking the PI response to calcium gating. The main components of the system (shown on the left) are a receptor (R), phospholipase C (Ph.C), a calcium gate and phosphatidylinositol (PI). The diagrams on the right illustrate three ways in which agonists might act to enhance calcium permeability. (a) The agonist occupies the receptor (R) which somehow activates phospholipase C (Ph.C) to hydrolyse PI to diglyceride (DG) and the gate opens. (b) The activated receptor acts directly on the gate. As the gate opens there are conformational changes in the surrounding bilayer which allows Ph.C. to hydrolyse PI as part of a closing mechanism. (c) The activated receptor stimulates Ph.C to hydrolyse PI to DG. The latter is rapidly phosphorylated to phosphatidic acid (PA) which functions as a calcium ionophore.

aggregated as occurs during capping, the rate of PI hydrolysis might exceed the rate of supply, causing depletion of this phospholipid within local domains, thus leading to a reduction in calcium signalling. Factors which alter membrane fluidity might also affect calcium signalling by regulating the supply of PI.

Nature of the calcium gate. A serious obstacle to our understanding of this signal pathway is our ignorance concerning the nature of the gate itself. The natural predilection is to consider that it is made up of an integral membrane protein(s) which opens in response to receptor activation. The hydrolysis of PI could then be associated either with the opening or closing of this calcium gate (Fig. 4a and b). In the latter case, PI might function in an analogous way to GTP in the adenylate cyclase system, i.e. it functions as an internal ligand whose presence is essential in order for the receptor to communicate with the calcium gate during an 'on-reaction'. The cleavage of PI resulting in a closing of the gate would then constitute an 'off-reaction'. Another possibility is that calcium is carried across the membrane by a phospholipid formed during the PI response. Studies on artificial membranes indicate that phosphatidate can function as a calcium ionophore. The diglyceride formed during the PI response is known to be rapidly phosphorylated to form phosphatidic acid which may then gate calcium[12,13] as shown in Fig. 4c. There is clearly a need to find out more about the precise nature of the effector system responsible for calcium gating across both outer and intracellular membranes.

The relationship between receptors and phospholipase C

Since phospholipase C appears to be a soluble enzyme[14] a problem arises concerning the way in which surface receptors are linked to this enzyme. In Fig. 4a and c, the receptor is shown communicating directly with the enzyme but there is no evidence for such an arrangment. Another possibility is that the receptor acts directly on the calcium gate (Fig. 4b) which on opening may make associated PI molecules more accessible to the underlying enzyme[11]. Irvine *et al.*[14] have made the interesting observation that phospholipase C degrades free PI very much more rapidly than PI which is inserted into a membrane. They consider that the activity of the enzyme may be controlled by the 'physicochemical state of its membrane substrates'.

Conclusion

The activity of most cells is regulated by a limited repertoire of second messengers. Translation of external signals into these internal signals entails a flow of information along a pathway comprising receptors, transducers and effector systems. For those receptors that exert stimulatory or inhibitory effects on the effector adenylate cyclase there is growing biochemical information on the role of GTP-regulatory proteins as transducers. Much less is known about the way in which agonists generate calcium signals.

All receptors that act through calcium have one thing in common, they all seem to stimulate the hydrolysis of phosphatidylinositol (PI). Care must be taken to divorce evidence that receptors are linked to the hydrolysis of PI from evidence that the latter is responsible for calcium gating. There seems to be little doubt that the PI response is an early event associated with the activation of a large number of different receptors which are also known to act through calcium. It is attractive to think that this hydrolysis of PI is responsible for generating a calcium signal as proposed by Michell *et al.*[7-9]. However, it is important to realize that evidence for this hypothesis is still somewhat circumstantial. Perhaps the most direct evidence presented so far comes from studies on an insect salivary gland where calcium gating could be varied by adjusting the level of PI in the membrane. However, many questions remain to be answered before this hypothesis can be fully accepted. Not the least of these concerns the nature of the effector system. In order to establish that PI hydrolysis is a sequential part of the signal pathway, it will be necessary to establish first the nature of the ionophore which allows calcium to cross the membrane. A related problem concerns the nature of intracellular calcium stores and how they might be influenced by a PI response presumably located in the surface membrane. It is clear that we are still a long way from understanding how agonists generate calcium signals but the possible involvement of PI hydrolysis offers what might turn out to be the vital first clue to this very important signal processing mechanism.

Reading list

1 Rodbell, M. (1980) *Nature (London)* 284, 17–22
2 Hirata, F., Strittmatter, W. J. and Axelrod, J. (1979) *Proc. Natl. Acad. Sci. U.S.A.* 76, 368–372
3 Sabol, S. L. and Nirenberg, M. (1979) *J. Biol. Chem.* 254, 1913–1920
4 Putney, J. W., Weiss, S. J., Leslie, B. A. and Marier, S. H. (1977) *J. Pharmacol. Exp. Ther.* 203, 144–155
5 Harper, J. F. and Brooker, G. (1979) *Mol. Pharm.* 13, 1048–1059
6 Hokin, M. R. and Hokin, L. E. (1954) *J. Biol. Chem.* 209, 549–558
7 Michell, R. H. (1979) *Trends Biochem. Sci.* 4, 128–131
8 Michell, R. H. (1979) in *Companion to Biochemistry* (Bull, A. T., Lagnado, J. R., Thomas, J. O. and Tipton, K. F., eds), Vol. 2, pp. 205–228, Longman, London
9 Michell, R. H., Jones, L. M. and Jafferji, S. S. (1977) *Biochem. Soc. Trans.* 5, 77–81
10 Fain, J. N. and Berridge, M. J. (1979) *Biochem. J.* 178, 45–58
11 Berridge, M. J. and Fain, J. N. (1979) *Biochem. J.* 178, 59–69
12 Salmon, D. M. and Honeyman, T. W. (1980) *Nature (London)* 284, 344–345
13 Putney, J. W., Weiss, S. J., Van de Walle, C. and Haddas, R. A. (1980) *Nature (London)* 284, 345–347
14 Irvine, R. F., Hemington, N. and Dawson, R. M. (1979) *Eur. J. Biochem.* 99, 525–530
15 Fain, J. N. and Garcia-Sainz, J. A. (1980) *Life Sci.* 26, 1183–1194

Why is phosphatidylinositol degraded in response to stimulation of certain receptors?

Robert H. Michell and Christopher J. Kirk

Department of Biochemistry, University of Birmingham, P.O. Box 363, Birmingham B15 2TT, U.K.

On arrival at the surface of a responsive cell most intercellular messages, including neurotransmitters, antigens and many hormones (other than steroids and thyroid hormones), are read and interpreted by receptors exposed on the exterior of the cell. But the cellular responses to such chemical messages frequently occur at intracellular sites (e.g. by modulation of the activity of contractile components or of cytoplasmic enzymes). Mechanisms must therefore exist for the the translation of these intercellular missives into a language understandable within the cell. In one such mechanism, receptor activation causes a change in the intracellular concentration of cyclic AMP, and we are at last within sight of molecular descriptions of three of the essential components; these are the receptor, a guanyl nucleotide-dependent protein that passes information on from the activated receptor, and the catalytic sub-unit responsible for cyclic AMP synthesis[1]. Knowledge of other systems is more scant, ranging from identification and isolation of the receptor without any profound understanding of mechanism (e.g. increased membrane sodium conductance triggered by the nicotinic cholinergic receptor) to almost complete ignorance (e.g. the actions of putative peptide neurotransmitters on target neurones in the CNS). A factor that contributes to our ignorance in this field is the frequency with which a single extracellular signal can serve multiple functions. For example, cells may translate the adrenaline message in three different ways, by using α_1, α_2 and β-receptors[1]; other ligands such as histamine, acetylcholine, opiate peptides and vasopressin can also be interpreted in disparate ways[1].

The various receptors that stimulate adenylate cyclase do so in a way that involves the sharing of a common pool of guanyl nucleotide-binding protein and adenylate cyclase[1], and it is to be hoped that out of the remaining diversity of ill-understood receptors will emerge additional families in each of which the members share a single mode of communication with the cell interior. Recognition of such families might, by analogy with the past history of progress towards an understanding of adenylate cyclase-coupled receptors, allow advances in the understanding of other widely distributed mechanisms to flow from detailed study of only one or two exemplars of each family.

One such family consists of a miscellany of receptors which share two intriguing characteristics. Firstly, their activation causes the breakdown of phosphatidylinositol (PI), a quantitively minor phospholipid, and this is usually followed by a compensatory increase in the rate of its synthesis (this set of events is often described as the 'PI response', see Fig. 1). Secondly, they control intracellular responses by somehow provoking an increase in the ionised calcium concentration of the cytosol compartment of the cell (a process we shall refer to as 'calcium mobilization'). Some of the receptors that evoke these responses are listed in Table I, and this association between PI breakdown and calcium mobilization has been discussed briefly both in *TIPS*[1] and in its sister journal, *TIBS*[2].

In this review we will attempt a brief critical analysis of an hypothesis, first published in 1975, which envisages that PI breakdown is a coupling reaction essential to the mechanism by which receptor activation leads to calcium mobilization (Fig. 1)[2-7]. The assessment of the hypothesis will be moved progressively into focus by explicit reference to a series of increasingly restrictive questions.

Are the receptors that stimulate PI breakdown and mobilize calcium always the same ones, and are they always distinct from those that control adenylate cyclase?

Some of the clearest answers to these questions come from recent studies with adrenergic ligands, vasopressin and 5-hydroxytryptamine, all of which can exert control over target cells either by mobilizing calcium or by regulating adenylate cyclase activity.

There are now four recognized classes of adrenergic receptors and between them they appear to call upon three different coupling mechanisms: both β_1 and β_2 receptors stimulate adenylate cyclase, α_2-receptors inhibit adenylate cyclase, and

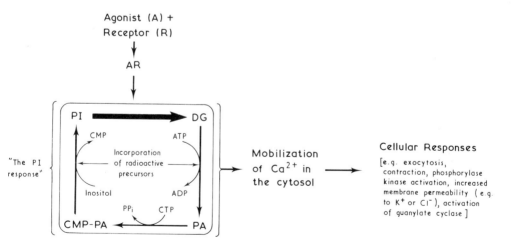

Fig. 1. A scheme illustrating the function suggested for PI breakdown as a coupling event in receptor-controlled mobilization of Ca^{2+} in the cytosol of many cells. The legend 'incorporation of radioactive precursors' within the cycle of the PI response refers only to incorporation of precursors into the phosphorylinositol headgroup of the lipid: during synthesis de novo the diacylglycerol backbone is synthesised by pathways not shown here[3]. Abbreviations: PI; phosphatidylinositol; DG, 1,2-diacylglycerol; PA, phosphatidate; CMP-PA, phosphatidyl-CMP (also known as CDP-diacylglycerol or CMP-phosphatidate).

α_1-receptors mobilize Ca^{2+} (Refs 1,5). All available evidence points to the conclusion that only α_1-receptors stimulate PI breakdown.

For vasopressin, 5-HT, and histamine, separate receptor populations have been identified which are responsible for the activation of adenylate cyclase and for the mobilization of Ca^{2+} within the cytosol. In each case, the receptors which evoke PI breakdown also control Ca^{2+} mobilization, and the activation of the adenylate cyclase-coupled receptors does not influence PI metabolism (Table I and Refs 1,4,7–9). Similarly, acetylcholine provokes PI breakdown via Ca^{2+}-mobilizing muscarinic receptors rather than through nicotinic receptors which evoke changes in Na^+-conductance[10]. Structure-activity relationships suggest that in the parotid gland, the same substance P receptors control both secretion (by mobilizing Ca^{2+}) and PI breakdown[11].

Recent evidence suggests that angiotensin II and muscarinic cholinergic stimuli, both of which exert intracellular effects via Ca^{2+}-mobilization, may also inhibit adenylate cyclase in some target tissues. It remains to be established whether, as with α_1- and α_2-receptors, there are two sub-sets of functionally distinct muscarinic and angiotensin II receptors.

Is PI breakdown a universal response to stimulation of calcium-mobilizing receptors?

There are few ligands whose PI responses have been studied in a substantial variety of target tissues, so it is often impossible to be certain how widespread is the response to a particular type of stimulus. Much the most extensive information comes from studies of the effects of α-adrenergic and muscarinic cholinergic stimuli. Such ligands produce PI responses in a wide variety of functionally diverse tissues (Table II) and, perhaps more important, we are not aware of any tissue in

which physiological or pharmacological studies have identified functional α_1-receptors or muscarinic cholinergic receptors and in which no PI response to these stimuli is seen.

Writing in the autumn of 1978, one of us said: 'I know of no well-defined cellular response which involves Ca^{2+} as a second messenger and in which there is no accompanying PI response'[2]. This statement stands two years later and it still seems probable that PI breakdown is a universal accompaniment of receptor-stimulated calcium mobilization, irrespective of the receptor or tissue being studied.

What is the relationship between receptor occupation and PI breakdown?

An early observation was that substantial stimulation of PI metabolism in the exocrine pancreas required much greater carbamylcholine or pancreozymin concentrations than did protein secretion, and this led to the inference that changed PI metabolism must be unrelated to the mechanisms that link receptor activation to secretion (see Refs 3,12). As we have seen above, however, all the available evidence suggests that the same receptors are responsible for evoking Ca^{2+}-mediated cellular responses and for stimulating PI breakdown. In order to reconcile these two observations, we suggested that PI breakdown might be a response that is relatively directly coupled to the activation of receptors, and that many 'classical' responses lie downstream of subsequent amplification steps in a coupling cascade (see Fig. 1)[10,12]. Stated in pharmacological terms, the proposal is that the PI response usually shows little or no 'receptor reserve', whereas other tissue responses often exhibit a large receptor reserve. Returning to the analogy of adenylate cyclase-coupled receptors, it is now clear that the efficiency of coupling between activated receptors and cyclase can vary from system to system, but that the curve describing activation of adeny-

TABLE I. Receptors whose stimulation provokes a 'PI response' (PI breakdown or enhanced PI labelling) in appropriate target tissues

Type of receptor	Tissues	References
Cholinergic (muscarinic)	Table II	
Adrenergic (α_1)	Table II	
Histamine (H_1)	Smooth muscle, brain	
5-Hydroxytryptamine (5-HT_1)	Smooth muscle, blowfly salivary gland, brain	1
Vasopressin (V_1)	Liver, smooth muscle	8,9
Pancreozymin	Pancreas, smooth muscle	
Substance P	Parotid, hypothalamus	11
Bombesin	Pancreas	
Angiotensin II	Liver	
Bradykinin	Endothelia, kidney	
Thyrotropin	Thyroid	
Nerve growth factor	Ganglia, pineal	
f-met-leu-phe, phagocyticable particles	Neutrophils	16,17
Secretagogues (antigen, etc.)	Mast cells	16
Thrombin, collagen, ADP	Platelets	18,19
Membrane depolarization	Smooth muscle, ganglia, synaptosomes	
Glucose	Islets of Langerhans	
Mitogenic stimuli (lectins, antisera, serum factors, transforming viruses etc.)	Various culture cells	

Further information on most of these observations can be found in Refs 1–6 and in the references cited above.

late cyclase by an agonist always falls between (or in the limiting case, on one of) the non-identical curves that describe receptor occupation and cellular response, respectively.

TABLE II. Tissues showing a PI response when exposed to adrenergic or cholinergic stimulation

Adrenergic: hepatocytes (α_1), adipocytes (α_1), pineal gland (α), brain (α, various regions), iris muscle (α), vas deferens (α), aorta (α), parotid, sublingual and submaxillary glands (α), eel gills (α?).

Cholinergic: brain and synaptosomes (musc., various regions), stellate and superior cervical ganglia (musc.), adrenal medulla (musc.), adenohypophysis (musc.), parotid, submaxillary, lacrimal and sweat glands (musc.), ileum, iris and vas deferens smooth muscles (musc.), avian salt gland (musc.), thyroid and pineal glands and fish electroplaque (?).

Where possible, a pharmacological assignment of the type of receptor responsible for the PI response is given. Information on most of these responses can be found in Refs 1, 3, 5 and 20 (for adrenergic responses) and in Refs 3, 4, 6, 10, 12, 13, 20 (for cholinergic responses).

There are now examples of PI responses whose dose-response curves fall close both to the ligand binding curve (e.g. parotid substance P receptors[11] and hepatic vasopressin V_1 receptors[8]) and to the physiological response curve (e.g. muscarinic cholinergic receptors in the parotid gland[13]; but compare with the rather different results elsewhere: see Refs in 6,10). On the basis of these results, it remains reasonable to consider that PI breakdown might always be more directly coupled to receptor occupation by agonists than are most cellular responses, and hence that it might be an amplification step on the pathway from activated receptor to classical cellular response (Fig. 1).

Is PI breakdown a response to calcium mobilization?

If our suggestion that ligand-stimulated PI breakdown is intrinsic to receptor-stimulated Ca^{2+}-mobilization is correct, *then it remains crucial to demonstrate un-equivocally that PI breakdown is not itself a consequence of such Ca^{2+}-mobilization.* Recent evidence has both reinforced this conclusion and put it seriously into question. A part of this confusion may arise from the failure of some workers in the field to acknowledge the essential difference between reactions whose occurrence is under the control of small changes in cytosol Ca^{2+} concentration and reactions which are simply dependent upon the presence of very low concentrations of the ion.

Before considering the available data, let us try to decide, by a comparison of dose-response curves, what would be the relative intracellular concentrations of Ca^{2+} needed for provoking typical cellular responses and PI breakdown if indeed they were both responses to a rise in cytosol Ca^{2+} (Refs 10,14). Where dose-response curves describing physiological responses and PI breakdown coincide, one would suspect that Ca^{2+} requirements would be the same. Where, by contrast, there is a receptor reserve with respect to the physiological response and the PI response curve lies closer to the receptor occupation curve, then a Ca^{2+}-mediated PI response would be triggered only at higher cytosol Ca^{2+} concentrations than those provoking the physiological response that displayed the receptor reserve.

One common way of attempting to determine the role of cytosol Ca^{2+} in the mediation of cellular responses is by gentle Ca^{2+} deprivation (e.g. by incubating tissues or cells briefly with Ca^{2+}-free medium or a medium containing a low concentration of a chelator such as EGTA). If the PI response was a result of a receptor-stimulated rise in cytosol Ca^{2+}, then it should be at least as sensitive to extracellular Ca^{2+}-deprivation as are classical cellular responses such as contraction and secretion. In systems that display a receptor reserve, a Ca^{2+}-mediated PI response would necessarily be more sensitive to Ca^{2+}

deprivation than would other cellular responses.

In most tissues in which the sensitivity to Ca^{2+} deprivation of the PI response has been compared with that of Ca^{2+}-mediated cellular responses the answer has been clear: the PI response is either insensitive to removal or mild chelation of extracellular Ca^{2+} or it is substantially less sensitive to such treatments than are other responses. Examples would include effects of muscarinic cholinergic stimuli (adrenal medulla, parotid, anterior pituitary, lacrimal gland), α-adrenergic stimuli (parotid, hepatocytes, lacrimal gland, pineal gland, vas deferens, iris muscle), 5-hydroxy-tryptamine (blowfly salivary gland), vasopressin (hepatocytes), angiotensin (hepatocytes) and substance P (parotid gland)[1-8]. Essentially similar studies may be performed by including, in the extracellular medium, drugs that retard receptor-stimulated Ca^{2+} influx into cells. In a detailed study of this type on ileum smooth muscle, the PI response proved insensitive to a variety of 'Ca^{2+}-antagonistic' drugs that inhibit the contractile responses to muscarinic cholinergic and H^1-histamine stimulation[14].

The possibility that Ca^{2+} may have a role in provoking PI breakdown has also been tested through the use of ionophores that admit Ca^{2+} to the cell and thence evoke Ca^{2+}-triggered responses. From the arguments presented above, it is clear that no useful conclusion can be drawn from experiments performed with ionophore concentrations that are only barely sufficient to evoke those cellular responses which display a receptor reserve, since failure to provoke PI breakdown in such cases might simply reflect insufficient Ca^{2+} mobilization in the cytosol. It is probable that this criticism is germane to some studies with ionophores. However, there are several tissues in which relatively high concentrations of ionophore fail to evoke PI breakdown (e.g. blowfly salivary gland, hepatocytes, parotid gland). [Although frequently undertaken, ionophore studies which simply analyse for changes in PI labelling are of limited value because: (a) Ca^{2+} may influence PI-synthetic enzymes without affecting PI degradation[1], and (b) Ca^{2+} can provoke the breakdown of inositol lipids other than PI, leading to PI labelling which is unrelated to prior breakdown of the lipid[2].]

There have, however, been recent reports suggesting a greater degree of involvement of Ca^{2+} in the triggering of PI breakdown and resynthesis than had previously been envisaged. (1) In synaptosomes the muscarinic PI labelling is reduced in low-calcium medium and can be abolished by Ca^{2+} chelators (see Ref. 15): unfortunately there was no well-defined Ca^{2+}-mediated 'physiological' response included in these studies for comparison. (2) Most stimuli which provoke Ca^{2+}-dependent mast cell secretion also provoke a PI labelling that is resistant to inhibition by Ca^{2+}-deprivation, but ATP^{4-} produces a PI labelling that is abolished by Ca^{2+} deprivation and appears to some extent dissociable from the secretory response (see Ref. 16). (3) fmet-leu-phe can produce a limited secretory response in polymorphonuclear leucocytes under conditions of Ca^{2+}-deprivation in which no PI breakdown or increase in PI labelling is seen[16,17]. (4) Whilst ionophore treatment does not cause PI breakdown in most tissues (e.g. parotid gland, blowfly salivary gland, anterior pituitary, mast cells, hepatocytes) there are at least two cells, namely platelets[18] and polymorphonuclear leucocytes[17], where Ca^{2+} admitted with an ionophore can provoke PI breakdown. In the platelet, however, there is recent evidence which suggests that ionophore-stimulated and thrombin-stimulated PI breakdown proceed by different mechanisms[19]. The three cells in which there seem likely to be Ca^{2+}-triggered mechanisms for PI breakdown are all related types that arise from stem cells in the bone marrow: whether dual mechanisms for provoking PI breakdown might also exist in cells other than the platelet (and its relatives?) is therefore an intriguing question.

Taken overall, this information suggests that a few tissues possess Ca^{2+}-regulated enzymic machinery for causing PI breakdown, but that *in the great majority of situations in which activation of receptors stimulates PI breakdown this happens by a mechanism in which Ca^{2+} is not an essential regulatory component* (the possibility remains that some step in this train of events may have a *requirement* for trace quantities of Ca^{2+}). The importance of the relatively small amount of recent information which casts some doubt on this conclusion is still difficult to assess, partly because much of it considers the labelling of PI in response to stimuli (an event that is usually, though not always, a secondary consequence of prior breakdown of PI). Detailed descriptions of the control of PI breakdown in these systems are now urgently needed.

Despite these doubts, the correlation between ligand-stimulated Ca^{2+} mobilization and PI breakdown is now far more impressive than when it was first noted[3]. The idea that PI breakdown is an essential coupling step in Ca^{2+} mobilization has received especially strong support from an experiment in which PI-depleted and 5-hydroxytryptamine-desensitized blowfly salivary gland was allowed to resynthesize PI and, maybe as a result, regained responsiveness to 5-hydroxytryptamine[2].

Ever since the association between ligand-stimulated PI breakdown and Ca^{2+} mobilization was first noted, an outstanding problem in the field has been to understand the mechanistic basis of this relationship. This is especially difficult since some ligands which stimulate PI breakdown are capable of mobilizing Ca^{2+} not only from the extracellular fluid but also from sources within the cell (Refs 1 and 2). A plausible, but disputed, idea as to how PI breakdown might bring about calcium mobilization has arisen from the realization that phosphatidate, which is synthesized from the diacylglycerol released during PI breakdown, might act as a natural calcium ionophore in stimulated tissue. This and other possible mechanisms are discussed in Refs 1 and 4.

A major problem remains the lack of any highly selective inhibitors of the phospholipase(s) C responsible for PI breakdown, which would, if this functional model is correct, necessarily antagonize the effects of all calcium-mobilizing ligands. These reagents, if they exist, must be poisons of unexplained action. Maybe they are sitting in dusty bottles in a drug company store or on a pharmacologist's reagent shelf – might it be yours? Perhaps such compounds have even been recognized as neurotoxins, since it seems likely that whatever understanding of receptor mechanisms we glean from studies of PI metabolism in peripheral tissues will ultimately prove equally useful to analysis of cell-cell communication in the nervous system[20].

Reading list

1 Berridge, M. J. (1980) *Trends Phamacol. Sci.* 1, 419–424
2 Michell, R. H. (1979) *Trends Biochem. Sci.* 4, 128–131
3 Michell, R. H. (1975) *Biochim. Biophys. Acta* 415, 81–147
4 Michell, R. H., Jafferji, S. S. and Jones, L. M. (1977) in *Function and Biosynthesis of Lipids* (Bazan, N. G., Brenner, R. R. and Giusto, N. M., eds), pp. 447–464, Plenum Press, New York
5 Jones, L. M. and Michell, R. H. (1978) *Biochem. Soc. Trans.* 6, 673–688
6 Jones, L. M., Cockcroft, S. S. and Michell, R. H. (1979) *Biochem. J.* 182, 669–676
7 Kirk, C. J., Billah, M. M., Jones, L. M. and Michell, R. H. (1980) in *Hormones and Cell Regulation* (Dumont, J. and Nunez, J., eds), Vol. 4, pp. 73–88, Elsevier/North Holland, Amsterdam
8 Kirk, C. J., Michell, R. H. and Hems, D. A. (1981) *Biochem. J.* 194, 155–165
9 Takhar, A. and Kirk, C. J. (1981) *Biochem. J.* 194, 167–172
10 Michell, R. H. (1980) in *Cellular Receptors for Hormones and Neurotransmitters* (Schulster, D. and Levitzki, A., eds), pp. 353–368, John Wiley, London
11 Hanley M. R., Lee, C. M., Jones, L. M. and Michell, R. H. (1980) *Mol. Pharmacol.* 18, 78–83
12 Michell, R. H., Jafferji, S. S. and Jones, L. M. (1976) *FEBS. Letters* 69, 1–5
13 Weiss, S. J. and Putney, J. W. *Biochem. J.* (in press)
14 Jafferji, S. S. and Michell, R. H. (1976) *Biochem. J.* 160, 163–169
15 Fisher, S. K. and Agranoff, B. W. (1980) *J. Neurochem.* 34, 1231–1240
16 Cockcroft, S., Bennett, J. P. and Gomperts, B. D. (1980) *FEBS. Letters* 110, 115–118
17 Cockcroft, S., Bennett, J. P. and Gomperts, B. D. *Nature (London)* (in press)
18 Rittenhouse-Simmons, S. and Deykin, D. in *Platelets in Biology and Pathology* (Gordon, J. L., ed.) (in press)
19 Lapetina, E. G., Billah, M. M. and Cuatrecasas, P. *J. Biol. Chem.* (in press)
20 Michell, R. H. *Neurosciences Res. Prog. Bull.* (in press)

Organization and transduction of peptide information

Robert Schwyzer

Department of Molecular Biology and Biophysics, Swiss Federal Institute of Technology Zürich (ETHZ), CH-8093 Zürich, Switzerland.

Introduction

In conjunction with pharmacological, biological, and physicochemical studies, the investigation of polypeptide hormones by chemical-synthetic means has, during the last two decades, led to great progress in our understanding of the anatomy of hormone molecules and of their mechanism of action on the molecular and cellular levels.

This has its practical consequences. Our anatomical knowledge allows us to prepare simpler molecules with greater potency, less side-reactions, or an altered spectrum of therapeutic properties (e.g. adrenocorticotropin partial sequences). Many problems, like oral application or prolonged action, however, remain to be solved in many cases.

The molecular mechanism of action on the target cells proceeds through receptors[1]. Much insight has been gained into receptor mechanisms and the structural requirements of receptors by circumstantial evidence, despite the fact that peptide hormone receptors still are, chemically, very elusive entities. They are extremely difficult to handle, because their activity is strongly dependent on their natural environment, the cellular lipid bilayer plasma membrane.

An understanding of receptor chemistry would be very desirable, however, because an increasing number of diseases seem to originate in molecular defects not of the hormones, but of the receptors.

The whole area has gained enormous importance because most of the well-known polypeptide hormones, as well as a host of recently discovered neuropeptides, have been found to act in the nervous system as neuromodulators and neurotransmitters. Detailed knowledge of the anatomy and mechanisms of action of these peptides is expected to lead to our better understanding of drug action in general and, especially, of three plagues of mankind: tolerance, dependence, and addiction. Perhaps even mental diseases could find their explanation on the molecular level.

The molecular anatomy of most of these peptides appears to follow the general rules that have been discovered in adrenocorticotropin (ACTH) and α-melanotropin (α-MSH). Molecular anatomy also forms the basis for the functional pleiotropy of these hormones and, consequently, of their genes.

Abbreviations are according to the recommendations of the IUPAC-IUB Joint Commission on Biochemical Nomenclature.

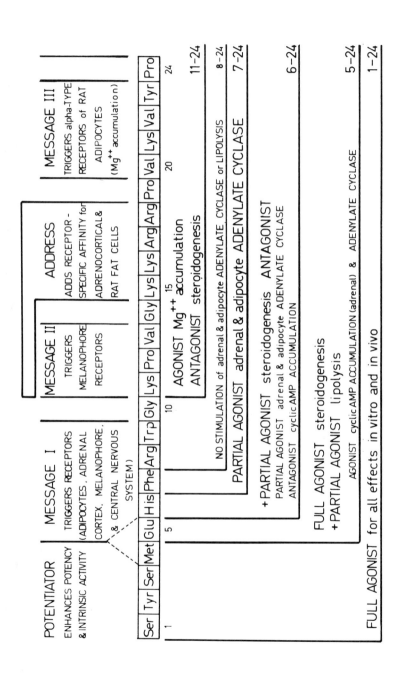

Fig. 1. An example of sychnologic information organization in a polypeptide hormone. The active site of ACTH(1–39) comprises the N-terminal tetrakosipeptide ACTH(1–24). The C-terminal part, ACTH(25–39), is not shown; it contains information about transport properties and stability in the blood and the species lable, but triggers no known ACTH action. Messages 1, 2, and 3 trigger different types of receptor. What is an address for certain receptors can contain a message for another. Potentiator and message 1 regions are slightly overlapping for different receptors: the potentiator for frog melanophores extends from 1 to 6, that for rat adrenocortical cells from 1 to 3. The intermediate region, Met-Glu is a message (besides message 2) for certain actions in the central nervous system (attentiveness).

In this short review, I shall try to give an account of the most recent achievements in this area with which I am connected, scientifically and emotionally.

The molecular anatomy and pleiotropic action of polypeptide hormones (see Ref. 2)

With their chain sequences of twenty different amino acids, peptides and proteins can be compared to sentences of our written language, with their twenty or more different letters. The arrangement of letters or amino acids can be either in the form of a clear text or of a secret code. The polypeptide hormones often correspond to the first, the protein hormones and the enzymes to the second arrangement.

Many polypeptide hormones, like ACTH, α-MSH, gastrin, glucagon, vasoactive intestinal peptide (VIP), substance P, parathyroid hormone, the enkephalins, and endorphins can be subdivided into at least two functionally different, structurally independent or slightly overlapping sequences of adjacent amino acids; one constituting the *active site* which triggers the receptor, the other an *auxiliary site* with multiple functions that are important for transport, species label, addressing to the correct target cell, or otherwise modifying the action, but without the capability of eliciting the hormonal response. The subdivision is chemically possible because adjacent sequences of amino acids (consecutive words) are responsible for the different parts of the total biological activity. As a consequence of the sequential proximity of the elements (amino acids) responsible for a certain function, this form of organization of information has been called *sychnologic* (from Greek συχνός, condensed, and λόγος, word).

Certain consecutive words have been given anthropomorphous names, like *address, message,* and *potentiator,* to describe the nature of their function. The best-studied example, that of the active site of ACTH (amino acids 1–24), is shown in Fig. 1. It can be seen that different tissues, cells, or receptors have different structural requirements and recognize different portions of the total sequence as their own message, potentiator, and address. The situation is similar to that of presenting the sequence WHOUSEDHAMSTERCELLS to Julius Caesar, Dante, Goethe, Mao Tsetung, or an Englishman. They will each recognize different regions as meaningful: SED (Latin), HO (Italian, Chinese), ER, AM, STER (German), and HOUSE, HAM (Englishman 1) or WHO USED HAMSTER CELLS (Englishman 2). We shall see that in adrenocortical cells there are, indeed, two Englishmen.

The secret code organization is found in enzymes and protein hormones like insulin. Here, the meaningful elements of the active site are scattered throughout the molecule and interdispersed in an unintelligible manner between the amino acids of the auxiliary regions. This form of information organization has been called *rhegnylogic* (from Greek ῥήγνυμι, tear apart, and λόγος, word). In this case, the auxiliary elements are responsible for a rather stable tertiary structure that unites the message elements in a well-defined manner to a topological element of the molecular surface called the active site. Different regions of this active site can be shown to have slightly different functions (one in insulin, for example, being responsible for the negative cooperativity of receptor interaction).

The main difference between the two types of organization will probably be found in molecular flexibility: the rhegnylogic type having fixed conformations interacting with the receptors; the sychnologic type being able to adapt flexibly to different receptor surfaces

(Fig. 2). Rhegnylogic hormones should therefore be expected to be more or less *monotropic* with respect to receptors (pleiotropy would result from the fact that very similar receptors would reside on different target cells, giving different responses). On the other hand, sychnologic hormones would be *pleiotropic* with respect to both function and receptors by virtue of their conformational adaptability and sychnologic organization.

These ideas remain to be investigated more closely.

Multiple receptors eliciting the same response

One of the greatest stimuli for research and understanding in the field of polypeptide hormone action during the last decade was undoubtedly provided by the discovery of cyclic AMP and the formulation of the cyclic AMP *second messenger concept* by Sutherland[3]. A veritable explosion of activity subsequently produced innumerable reports demonstrating effects of almost all known peptide and protein agonists on cyclic AMP production – either stimulatory or inhibitory – in their target cells.

The sheer beauty of the second messenger concept,

Hormone → cyclic AMP → Response

is equalled only by the so-called 'central dogma' of molecular genetics,

DNA → RNA → Protein.

Its appealing qualities have, indeed, also resulted in a somewhat dogmatic attitude of many scientists. Criticisms were often viewed with more than a

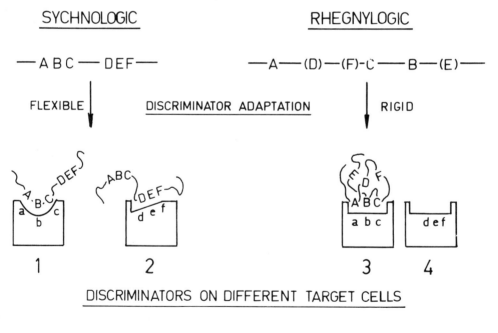

Fig. 2. Hypothetical adaptation of the flexible sychnologic hormones and the rigid rhegnylogic hormones to different receptors (discriminators). Sychnologic hormones can adapt to different discriminator surfaces with different sequential regions; rigid proteins only to one (3, not 4). The second case also describes the binding of a substrate (3) to an enzyme active site (ABC). With respect to receptor types, pleiotropic and monotropic actions can be distinguished.

normal and healthy amount of suspicion by referees and editors.

The work on ACTH especially (Fig. 1) and its actions on adrenocortical cells and adipocytes to produce, respectively, steroidogenesis and lipolysis as well as stimulation of adenylate cyclase (the enzyme producing cyclic AMP) led to the suspicion that the former responses could not simply be the consequences of enhanced cyclic AMP concentration within the cell. It was postulated that ACTH must, in addition to adenylate cyclase, also activate other receptor mechanisms[4-6].

The two main suggestions were that an acute action of ACTH at low physiological concentrations could stimulate steroidogenesis and lipolysis either by a pathway completely independent of cyclic AMP (dual receptor concept) or through a mechanism utilizing pre-existing background amounts of cyclic AMP (compartment guidance concept)[7].

With the help of ACTH (6–24) in its competitive antagonist mode (Fig. 1), it was recently demonstrated that at least two populations of steroidogenic receptor with two significantly different affinities for this antagonist exist in the rat adrenocortical cell (Fig. 3)[8]. ACTH triggers corticosterone production at concentrations somewhat lower than those needed to trigger cyclic AMP synthesis. With ACTH(5–24) (Fig. 1), however, this discrepancy is much more pronounced: steroidogenesis reaches a maximal rate at concentrations which elicit no discernible cyclic AMP production (the latter activity is stimulated only at roughly 1000 times greater agonist concentrations). The inhibition of the steroidogenic action of ACTH(5–24) also needs about 1000 times lower concentrations of ACTH(6–24) than the inhibition of steroidogenesis produced by ACTH (1–39).

Fig. 3 shows the interpretation of these findings; ACTH reacts with two different receptors on or in the rat adrenocortical cell membrane. Activation of receptor A produces steroidogenesis through cyclic AMP production; activation of receptor B produces steroidogenesis through some other mechanism (calcium redistribution, cyclic GMP production or utilization of background amounts of cyclic AMP?).

ACTH(5–24) contains at its N-terminus only the message for activating receptor B (Glu-His-Phe-Arg-Trp-); this message can stimulate receptor A only at very high concentrations. Receptor A needs more and perhaps different information than receptor B for normal stimulation. This additional information is probably provided by the methionine residue.

Thus the idea of two receptors for one and the same response and of receptor-based pleiotropy (two Englishmen!) for certain peptide hormones will have to be taken into serious account for future work.

ACTH spans artificial lipid bilayer membranes

Allegedly, peptide hormones do not penetrate the target cell membrane. Instead, they react reversibly with the cell–external parts of hypothetical receptors (the discriminators) which are believed to be integral proteins. This notion has developed into a kind of 'dogma'. Is it really true? Again, it is ACTH which might give us insight into the truth and the limitations of this concept.

Recently, a new method was developed in my laboratory with which the asymmetric attachment of charged molecules (e.g. peptide hormones) to planar lipid bilayer membranes (BLM) can be measured for the first time with high sensitivity and precision. The method is called *capacitance minimization*[9].

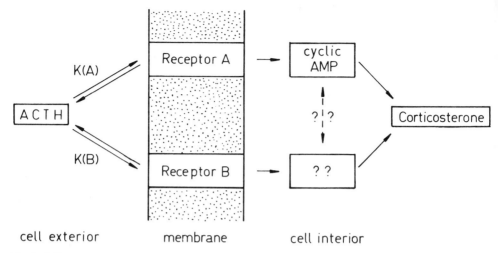

Fig. 3. A diagrammatic representation of the mechanism of ACTH action on isolated adrenocortical cells of rats.

A charge adsorbed to one side of a BLM separating two electrolyte compartments will set up an electric field across the interior of the membrane, resulting in a compressive force which causes the bilayer to thin, thereby increasing its electrical capacitance. By applying an external voltage to the aqueous solutions the internal field can be adjusted to zero, thus reducing the membrane capacitance to its minimum value. This externally applied voltage is called the capacitance minimization potential and is equal to the difference of surface potentials on the two sides of the membrane.

The application of this method to the study of ACTH(1–24)/BLM interactions has revealed an entirely unexpected feature (Fig. 4). When added to one side of a BLM (cis), ACTH(1–24) actually becomes adsorbed to the surface of the membrane in a reversible and eventually saturable fashion, equilibrium A (the adsorbed molecules tend to repel other incoming molecules and the net repulsion increases with the number of adsorbed charges). This adsorption can be demonstrated with the capacitance decrease caused by a shielding of the adsorbed charges with pH and ionic strength

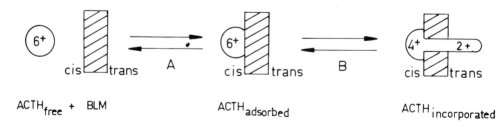

Fig. 4. The interaction of ACTH(1–24) with artificial lipid bilayer membranes (BLM).

changes on the *cis* side of the membrane, but not on the *trans* side. Equilibrium A is reached very rapidly (diffusion control?) with a velocity that is not measurable by our method.

For simply adsorbed peptides, like the bee-venom peptide melittin, this situation persists. Not so with ACTH(1–24). The capacitance slowly becomes susceptible to pH and ionic strength influences on the *trans* side.

It can be shown that this is the result of an ACTH/BLM interaction in which the hormone spans the membrane and two charges are exposed on the *trans* side. The interaction is reversible (equilibrium B) with a time constant of a few minutes. At 10^{-5}M ACTH(1–24) and neutral membranes, the ratio at equilibrium is about 2 molecules adsorbed to 1 molecule incorporated.

Although this point has not been proved, we have good reason to believe that the address region of ACTH(1–24) with 4 accumulated positive charges (-Lys-Lys-Arg-Arg-) remains on the *cis* side, and that it is the message that penetrates the membrane. This situation is comparable to that of the 'signal' or 'leader' peptides which aid the membrane transgression of excretory proteins during their synthesis.

It also has its consequences for our understanding of ACTH receptors (and other peptide hormone receptors?). With membranes containing about 5% of a negatively charged lipid, the ACTH interaction can easily be detected at 10^{-9}M ACTH(1–24) and probably much less. If the target cell membrane behaves similarly, then ACTH could react very quickly and at low concentrations with discriminators exposed near the outer surface; it could also react with discriminators *inside* the cell membrane or even with receptors on the cytoplasmic surface or in the cytoplasm.

It will be very interesting to see whether similar BLM interactions also exist for other polypeptide hormones and where the receptor discriminators are actually located. Our findings could also explain recent observations of Bennett and McMartin that injected ACTH is rapidly cleared by tissues that are not known to contain ACTH receptors, like skin, muscle, and intestines, from whence the hormone may re-enter the circulation again.

Conclusions and questions

The sychological organization of information is a distinctive feature of many peptide agonists (hormones and neuropeptides). Together with the flexibility of such molecules it allows for a great pleiotropy of receptor interactions and responses. It is demonstrated that two steroidogenically responsive receptors for ACTH exist and that only one is connected with the production of cyclic AMP. Do other peptide agonists also have multiple paths for eliciting their responses? Finally, very specific interactions of ACTH with 'receptorless' artificial lipid bilayer membranes have been discovered. They allow the hormone to span the membrane and to expose its message on the side opposite to that of application. Could this mean that receptors might be located in the interior of the membrane or even on its cytoplasmic side?

Acknowledgements

This work was supported by grants from the Swiss National Science Foundation and the ETHZ.

Reading list

1 Ariëns, E. J. and Simonis, A. M. (1964) *J. Pharm. Pharmacol.* 16, 137–157
2 Schwyzer, R. (1977) *Ann. N.Y. Acad. Sci.* 297, 3–26

3 Sutherland, E. W. (1972) *Science* 177, 401–408

4 Ramachandran, J. and Moyle, W. R. (1977) in *Endocrinology* (James, V. H. T. ed.), Vol. 1, pp. 520–525, Excerpta Medica, Amsterdam and Oxford

5 Seelig, S., Lindley, B. D. and Sayers, G. (1975) *Methods Enzymol.* 39, 347–359

6 Schwyzer, R. (1974) *Pure Appl. Chem.* 37, 299–314

7 Schulster, D. and Schwyzer, R. (1980) in *Cellular Receptors for Hormones and Neurotransmitters* (Schulster, D. and Levitzki, A. eds), pp. 197–217, John Wiley & Sons Ltd, London and New York

8 Bristow, A. F., Gleed, C., Fauchère, J. L., Schwyzer, R. and Schulster, D. (1980) *Biochem. J.* 186, 599–603

9 Schoch, P., Sargent, D. and Schwyzer, R. (1979) *J. Membrane Biol.* 46, 71–89

Angiotensin receptors and angiotensinase activity in vascular tissue

Djuro Palaic

Department of Pharmacology, University of Montreal, P.O.B. 6128 Montreal, P.Q., Canada H3C 3J7

Among many causes of hypertension, the renal cause appears to be best understood. The role of kidney in blood pressure regulation was recognized already in the last century when Tigerstedt and Bergmann[1] described a pressor substance from kidney which they called renin. Page and Braun Menandez[2] separately have found that renin is a proteolytic enzyme which acts on a substrate in plasma giving rise to a polypeptide named angiotensin. Actually, the product of renin action is a decapeptide angiotensin I which is converted to angiotensin II, an octapeptide. Angiotensin II, or simply angiotensin, is the most powerful natural pressor substance known today.

In this article I intend to discuss the mechanisms of action of angiotensin on vascular tissue. It is the present day belief that all endogenous hormones act on target tissue through their own receptors. These protein entities recognize the hormone and transmit the message to the 'effector' machinery which in turn brings about the physiological response. The interest in angiotensin receptors in vascular tissue was preceded by rather extensive studies of the indirect actions of angiotensin on the sympathetic nervous system. These studies done during the 'sixties and early 'seventies have shown that angiotensin is capable of stimulating adrenergic nerves by inhibition of noradrenaline uptake, stimulation of its release and through stimulation of brain centres involved in blood pressure control.

The lack of information on specific angiotensin receptors in vascular tissue was caused by inadequate and insensitive techniques and by failure to find a specific antagonist for angiotensin. In the early 'seventies these two obstacles were overcome by introduction of labelled angiotensin with high specific activity and by successful synthesis of a series of angiotensin analogs showing antagonistic action to angiotensin.

The characterization of vascular angiotensin receptors *in vitro* was done simultaneously by Phillippe Meyer and his group[3] and by us. The detailed study of angiotensin receptors in vascular tissue revealed that they are localized at the cell membrane level. They can be saturated with appropriate doses of angiotensin. The equilibrium of binding is achieved in a period of time compatible with the onset of physiological response. The receptor binding is reversible and the dissociation constant correlates well with *in vitro* tissue response. This binding can be effectively blocked by angiotensin antagonists but not with bradykinin or angiotensin I[4]. Ion strength and pH of the medium greatly influence the binding of angiotensin to receptors. 'High energy' compounds like ATP and GTP seem not to have any effect. Enzymatic treatment with trypsin and pepsin double the bind-

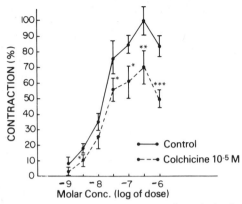

Fig. 1. Dose–response curve to angiotensin in the presence of colchicine[9].

ing while phospholipase A and C do not exert any action.

In 1971 we observed that isolated guinea-pig aorta made tachyphylactic to maximal doses of angiotensin would bind more labelled angiotensin than control[5]. This surprising finding led us to postulate that the number of angiotensin receptors in vascular tissue is not constant. At the time we suggested that "induction of new binding sites may possibly occur by *de novo* synthesis or by allosteric conformational changes of cell membrane proteins, which render them capable of binding more angiotensin, the supply of agonist being the determining factor in these processes". In today's jargon this would appropriately be called 'up regulation' of receptors. This hypothesis was further developed and documented by Meyer and his group[3]. These authors found an increase of angiotensin receptors *in vivo* after nephrectomy. This was due to lowering of endogenous plasma concentrations of angiotensin due to elimination of the major source of renin production and positive sodium balance. Sodium balance appears to be the most specific regulator of vascular tissue sensitivity to angiotensin. An increase in sodium balance results in increase of sensitivity while negative sodium balance diminishes the sensitivity. In nephrectomized

animals, while the number of angiotensin receptors rises there is no change in affinity of receptors for angiotensin.

Our recent studies reveal that another specific regulator of vascular angiotensin receptors are the tissue angiotensinases[6]. For some time it has been recognized that the metabolic fate of circulating angiotensin depends mostly on tissue angiotensinases and not on plasma angiotensinase. Angiotensinases make a mixed bag of peptidases. The one which attacks the N-terminal of peptide was named α-aminopeptidase (angiotensinase A); the one which acts on C-terminal was called carboxypeptidase (angiotensinase C); and the one which inactivates peptide fragments was called endopeptidase (angiotensinase B). All these enzymes have not been identified in vascular tissue. One can speculate that aminopeptidase and carboxypeptidase are there since the N-terminal and C-terminal of angiotensin are indispensible for contractile effect on vascular smooth muscle.

Our work on subcellular distribution of angiotensinases in vascular tissue showed that the localization of the enzymes plays a very important role[7]. One finds angiotensinase activity in the cell membrane fraction together with the angiotensin receptors. This enzyme is thermolabile since heat treatment at 47°C for 20 min destroys the enzyme activity while the binding to receptors does not change. Heat treatment increased the maximal response of tissue and accelerated the onset of tachyphylaxis to angiotensin[8]. Still, the bulk of angiotensinase activity localized in the cytosol fraction. It is apparent that intracellular angiotensinases represent a non-differentiated group of peptidases whose role in angiotensin action on vascular tissue is not understood.

In order to prove the importance of cell membrane angiotensinase in receptor operation and regulation we employed specific angiotensinase inhibitors, epsilon-amino-*n*-caproic acid and EDTA[6]. These compounds blocked cell membrane angiotensinase and increased the number of angiotensin receptors in cell membrane fraction. This resulted in an increased maximal response of vascular tissue to angiotensin while the apparent affinity of receptors for angiotensin remained unchanged. From these results we concluded that angiotensinase is involved in activation or 'uncovering' of additional angiotensin receptors. The diminished degradation of peptide by angiotensinase would increase its availability at the receptor site. Thus, these two factors:—increased number of receptors and decreased degradation of angiotensin, may indeed explain the increased maximal response to angiotensin after angiotensinase blockade by angiotensinase inhibitors.

Our work on angiotensin tachyphylaxis, work with angiotensinase inhibitors, and the results of Meyer's group clearly point out that angiotensin receptors are not rigid entities in cell membrane but rather mobile macromolecules in constant flux regulated by the supply of agonist and angiotensinase activity. Like any membrane protein the angiotensin receptor must be synthesized intracellularly and delivered to the cell membrane. Drugs which block protein synthesis, like cycloheximide, or interrupt the flow of intracellular proteins, like colchicine, do indeed block the vascular response to angiotensin (Fig. 1).

The possibilities of attacking the angiotensin–renin system by pharmacological means and of bringing about a change in sensitivity of vascular tissue exist at different levels. One can block renin and converting enzyme activity with specific agents. Recently, a specific inhibitor of angiotensin converting enzyme was introduced as an antihypertensive agent. Specific antagonists of angiotensin have already been known for some time, all of them being angiotensin analogs with very limited clinical use. The manipulation of vascular response to angiotensin is also possible through the change of sodium load and other factors. However, most of these changes are reversible and of short duration. One wonders whether a long-term or permanent change of receptors would be of greater therapeutic significance. One possible way of doing it is immunological manipulation of angiotensin receptors. The angiotensin receptor can be solubilized from the cell membrane by deoxycholic acid. This procedure successfully separates the receptor protein from angiotensinase[10] (Fig. 2). However, the pure angiotensin receptors are still not isolated. This work is hampered by the

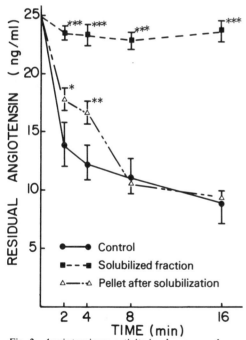

Fig. 2. *Angiotensinase activity in plasma membrane fraction before and after solubilization with 7-deoxycholic acid (0.010%). The pressor effect of residual angiotensin after incubation with different fractions was measured on rat blood pressure preparation. 0.25 μM angiotensin was incubated with 100-150 μg of protein/ml. *p <0.05; **p <0.01; ***p <0.001.*

relative scarcity of angiotensin receptors in vascular tissue. However, it is only a matter of time before this technical obstacle is removed, thus opening a new era of research in this field.

Reading list

1. Tigerstedt, R. and Bergman, P. G. (1898) *Skand. Arch. Physiol.* 8, 223–233.
2. Page, G. H. (1968) In: I. H. Page and J. W. McCubbin (eds) *Renal Hypertension,* Year Book Medical Publishers Inc., Chicago, pp. 391–396.
3. Devynck, M. A. and Meyer, P. (1976) *Am. J. Med.* 61, 758–767.
4. LeMorvan, P. and Palaic, Dj. (1975) *J. Pharmacol. Exp. Ther.* 195, 167–175.
5. Palaic, Dj. and LeMorvan, P. (1971) *J. Pharmacol. Exp. Ther.* 179, 522–531.
6. Farruggia, P., Sachs, S. and Palaic, Dj. (1979) *Mol. Pharmacol.* 15, 525–530.
7. LeMorvan, P., Palaic, Dj. and Ferguson, D. (1977) *Can. J. Physiol. Pharmacol.* 55, 652–657.
8. LeMorvan, P. and Palaic, Dj. (1972) *Can. J. Physiol. Pharmacol.* 50, 498–502.
9. Palaic, Dj., Khullar, H. and Farrugia, P. *J. Cardiovas. Pharmacol.* (in press).
10. Palaic, Dj. and Farrugia, P. (1979) *Clin. Exp. Hypertension* (in press).

Peripheral opiate receptor mechanisms

John Hughes

Imperial College of Science and Technology, Department of Biochemistry, Imperial Institute Road, London SW7 2AZ, U.K.

Introduction

The discovery of endogenous opioid peptides in the pituitary and brain owed much to the exploitation of knowledge gained about the actions of opiates in peripheral organs such as the guinea-pig ileum and mouse vas deferens. It seems therefore apposite that studies on the endorphins and enkephalins have given new insights into the nature and role of opiate receptor mechanisms in peripheral tissues. It was perhaps not surprising but it was certainly exciting when, having studied the distribution of enkephalins in the brain, Terry Smith, Hans Kosterlitz and myself[1] were also able to demonstrate significant quantities of enkephalin-like peptides in the gastro-intestinal tract, autonomic nervous structures and other tissues. A series of immunohistochemical studies confirmed and extended these observations[2,3] and it is now clear that enkephalin-like and β-endorphin-like immunoreactivity is widely distributed in the periphery. Consequent upon these discoveries are various reports suggesting the much wider distribution of opiate receptors than was hitherto suspected. It is premature to draw any conclusions concerning the physiological functions of peripheral opiate receptor systems but this seems an appropriate time in which to attempt to correlate the available facts and hypotheses.

Peripheral opiate receptors

Opiate receptors can be identified and located either by classical pharmacological techniques involving agonist modulation of physiological events and the use of stereoselective antagonists, or by receptor binding techniques utilising radiolabelled ligands. The latter method has only been successful at relatively few peripheral sites such as the guinea-pig and rabbit ileum, mouse vas deferens and adrenal medulla whereas pharmacological assays suggest a much wider distribution (Tables I and II). However at the present time the classification and indeed the definition of opiate receptors is in some disarray. It is quite clear from numerous receptor binding and parallel bioassay studies that multiple types of opiate receptors exist. These receptors have been given a variety of Greek letter prefixes, the definition and attribution of the letters varying from laboratory to laboratory depending on the particular techniques employed. At the present time mu- and delta-receptors are the most widely accepted subclassifications (Table III) since the evidence includes a distinct agonist potency series at each receptor in bioassay and ligand receptor displacement assays, different pA2 values for naloxone against putative mu- and delta-receptor agonists in delta-receptor 'enriched' preparations and more recently evidence from cross-protection and tolerance studies. Robson and Kosterlitz[4] were able to show that phenoxybenzamine irreversibly inhibits the receptor binding of [^{3}H]dihydromorphine and of [^{3}H](D-Ala$_2$-D-Leu$_5$)-

enkephalin. Prior incubation with cold enkephalin protected the enkephalin binding sites but not the dihydromorphine binding sites whereas cold dihydromorphine protected the dihydromorphine (mu-receptor) binding sites but not the enkephalin (delta-receptor) sites. Similarly Albert Herz[5] and his colleagues have shown that chronic treatment of mice with sufentanyl (mu-agonist) rendered the mouse vas deferens insensitive to other mu-agonists, but had little effect on the delta agonist (D-Ala$_2$-D-Leu$_5$)-enkephalin. The reverse occurring when the animals were chronically treated with the delta agonist. This latter type of analysis has been used also to support the idea that dynorphin receptors in the mouse vas deferens and β-endorphin receptors in the rat vas deferens are distinct from other mu-

TABLE I. Inhibition of neurotransmitter release by opioids at various peripheral sites

Site	Neurotransmitter	Predominant opiate receptor
Guinea-pig ileum*		mu
Rabbit ileum*		(delta)
Mouse ileum	Acetylcholine	(delta)
Rabbit atria		(mu)
Sympathetic ganglia*		(delta)
Cat nictitating membrane*		delta
Mouse vas deferens*		delta
Rat vas deferens	Noradrenaline	(mu)
Rabbit vas deferens		(mu)
Rabbit ear artery		(delta)
Splenic nerve*		(delta)
Guinea-pig taenia caecum	Unidentified inhibitory transmitter	?
Rat colon		

Note that neurotransmitter release has only been determined directly at those sites marked*. In only a few cases has sufficient pharmacological characterisation been carried out to determine the predominant receptor type; those designations in brackets are based on incomplete data or refer to receptors that do not fit the mu/delta classification completely.

TABLE II. Peripheral effects of opiates or opioid peptides

Site	Opiate mediated effect
Rat stomach	Inhibit somatostatin release
Rat jejunum	Inhibit PGE$_1$ stimulated adenyl cyclase activity and water transfer
Dog exocrine pancreas	Inhibit bicarbonate, water and enzyme secretion evoked by secretion or duodenal acidification
Dog endocrine pancreas	Inhibit somatostatin release with subsequent increase in insulin and glucagon secretion
Adrenal cortex	Increases corticosterone production
Adrenal medulla	Inhibit catecholamine release
Fat cells	Increases phosphofructokinase activity
	Decreases phosphohexoseisomerase activity
	Increases lipolysis (not blocked by naloxone)
Kidney	Decreases ornithine decarboxylase activity
Neuroblastoma/glioma cells	Inhibit adenyl cyclase
	Decreases AMP-dependent protein kinases
	Decreases ornithine decarboxylase activity
	Decreases synthesis of gangliosides and membrane glycoproteins.

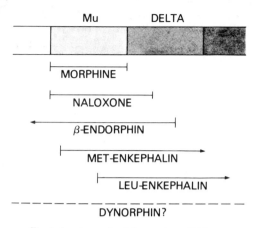

Fig. 1. Spectrum of opiate receptor activities.

and delta-receptors. However this type of analysis does not define a receptor type and caution should be exercised in interpretation. For example it is not clear whether the 'desensitization' is primarily a receptor or post-receptor directed effect and the results are only likely to be clear cut where there are enormous potency differences between ligands as there are in the mouse and rat vas deferens.

The evidence for a third opiate (kappa) receptor is less conclusive although kappa agonists such as ethylketocyclazocine are poorly antagonised by opiate antagonists in the guinea-pig ileum and mouse vas deferens. However binding studies indicate that mu-receptor ligands compete well for [³H]ethylketocyclazocine (EKC) high affinity binding sites and vice versa[6]. At present the proposed kappa receptor can only be defined as being somewhat less sensitive to naloxone and to interact with ligands that do not appear to substitute for mu-agonists in morphine dependent animals. There is a low affinity binding site for EKC that is insensitive to mu- and delta-ligands but the relevance of this is yet to be determined.

There are also examples of peripheral receptors that do not fit the above classifications. In the rabbit ear artery the enkephalin mediated inhibition of the noradrenergic contraction cannot be mimicked by mu-agonists. In the rat and rabbit vas deferens β-endorphin is many times more potent than other mu- or delta-agonists and yet β-endorphin is readily antagonised by naloxone albeit at some ten times the concentration required for other mu- receptor sites. These may be examples of extreme tissue/receptor selectivity since a similar separation of mu- and delta-receptors occurs in the vasa from different strains of mice. There are also responses which are not blocked by naloxone; thus a lipolytic effect of β-endorphin, and to a much lesser extent other opioids, has been demonstrated in rabbit adipocytes and this effect is mimicked by naloxone[7].

At present we are left with the opiate receptor defined in terms of that moiety which mediates a pharmacological response which can be stereoselectively antagonised by naloxone. Thus mu- and delta-receptors are opiate by definition, although this may be misleading since morphine and many of its cogeners are not delta-agonists. The differentiation between these receptors may be somewhat analagous to alpha₁ and alpha₂ adrenoceptors. If, as in Fig. 1, we represent the opiate receptors as a continuum, although they might be quite separate macromolecules, then we can represent the ligand specificity as shown. Thus extremes of activity or tissue selectivity may be found in which our present antagonists are ineffective and cannot be classified as opiate. Further definition of the receptor types will be only determined unequivocally by the development of more specific antagonists.

The apparent plurality of receptors is matched by a proliferation of putative endogenous ligands (Table IV). However that is not to say that this throws any further light on the classification of receptor types. At the present time it is impossible to assign a particular physiological function on the basis of ligand affinity for a specific receptor. A comparison with the catecholamines is instructive in this respect, for example the relative potency of adrenaline to noradrenaline gives no indication of the respective roles of these amines in any particular physiological situ-

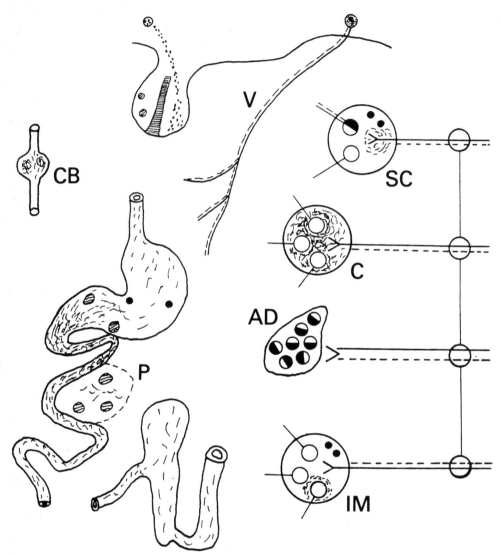

Fig. 2. Distribution of peripheral opioid peptides. Enkephalin-like immunoreactive (ELI) neurones (- - - -) apparently arise in the spinal cord and innervate ganglion cells and adrenal medulla via the ventral roots, ELI-neurones are also present in the vagus nerve (V). It is not certain whether the ELI is present in cholinergic or separate neurones. Various patterns of ELI (𝔫 𝔰) are seen in ganglia varying from patches to dense innervation. ELI in the gut is mainly confined to neurones which probably arise from intrinsic ganglia (not shown). ELI cell bodies and chromaffin granules (⬤) are also seen in sympathetic ganglia and adrenal medulla where in each case there is co-storage with noradrenaline. A number of SIF-cells (•) have also been reported to contain ELI. β-endorphin-like immunoreactivity (≡) is seen in cells of the intermediate lobe of the pituitary and to a much lesser extent in the anterior lobe; other immunoreactive cells are seen in the pancreas (P) and gastrointestinal tract. Dynorphin (1–13)-immunoreactivity (•∴•) is present in the posterior lobe of the pituitary. SC, C, IM – sympathetic ganglia, AD – adrenal medulla, CB – carotid body. Note that these distributions probably vary markedly with species and can only been regarded as approximations at present.

ation. In this particular case the generally less potent noradrenaline subserves a neurotransmitter function whilst adrenaline has a hormonal role. Taking this comparison further I would consider it extremely unlikely that one receptor type interacts specifically with one ligand. Thus at some mu-receptor specific sites β-endorphin may act normally as a mediator whilst at other mu-sites methionine-enkephalin may be the natural mediator.

Peripheral endogenous opioid peptides

The available evidence suggests that the peripheral opioid peptides could act as endocrine (from pituitary and adrenal glands), apocrine (in stomach and pancreas) or neuronal hormones (sympathetic ganglia and nervous plexi of gut). Fig. 2 illustrates the postulated distribution of β-endorphin-, dynorphin- and enkephalin-like peptides in various peripheral tissues. Complete chemical characterisation is lacking for the peptides at most of these sites. In the pituitary it has been estimated that less than 10% of the endorphin immunoreactivity is due to authentic β-endorphin (other known forms include LPH_{61-87}, N-acetyl-β-endorphin and N-acetyl-LPH_{61-87}), whilst less than 5% of the adrenal enkephalin-like peptides are authentic enkephalin. Some of these undetermined structures may just be precursors or storage depots for the opioid peptides but is is possible that multiple types of opioid peptide are released together and either have a concerted or dispersed mode of action.

Physiological roles of peripheral opioids

Inspection of tables I and II gives an idea of the possible tissues and physiological processes which may act as targets for peripherally released opioid peptides. In general these phenomena have been little investigated with respect to their physiological significance. It is not necessarily true that the presence of opiate receptors indicates a physiological role for the receptor or an endogenous opioid ligand. However it is interesting to speculate on possible physiological functions of the various systems.

Circulating opioid peptides

Opioid peptides released from the pituitary may modulate biochemical events at such diverse sites as the adrenal cortex, kidney, pancreas, sex organs and fat cells. The available evidence suggests that β-endorphin or like peptides may be involved in a homeostatic response to stress since lipotropin-endorphin peptides are released under such conditions in parallel with adrenocorticotropin. There are several reports that naloxone treatment can exacerbate stress responses thus providing additional support for an ameliorative role for opioid peptides in stressful conditions. For example naloxone enhances oedema formation (elicited by carrageenin) and markedly increases gastric erosions in cold stressed rats[8]. There are also reports that the normally low β-endorphin levels in blood are markedly elevated during birth in both mother and child. Dynorphin[9], which has yet to be fully sequenced, is mainly concentrated in the posterior lobe of the pituitary gland. Thus this peptide may be involved in responses to changes in water and salt balance, and possibly in processes associated with birth and suckling.

The possible roles of enkephalin-like peptides in the adrenal medulla are intriguing. In certain animals, such as the cow, there are extremely high levels of these peptides in the adrenal chromaffin cells where they are stored with the catecholamines. Recent studies by H-Y. T. Yang and E. Costa[10] and by O. H. Viveros and colleagues[11] indicate that there are opiate receptors in adrenal membranes, that the opioid peptides are released in parallel with the catecholamines by such

TABLE III. Opiate receptor classification

Designation	Potency series	Naloxone Ke	Sodium 'shift'
mu	BE > MOR > ME > LE	1–3 nM	20–100
delta	LE > ME > BE >> MOR	20–40 nM	5–20
kappa	undefined[1]	6–12 nM	<5
sigma	undefined[2]	?	<5

Note that although β-endorphin (BE) is a very potent mu-agonist it requires 10 times more naloxone to reverse its action in the mouse vas deferens than morphine (MOR). Also the potent delta agonist methionine-enkephalin (ME) and leucine-enkephalin (LE) behave as typical mu-agonists in the guinea-pig ileum with respect to naloxone antagonism.

The sodium shift is a measure of the decrease in receptor binding that occurs when measured in sodium containing media compared to sodium free. [1]Kappa agonists are represented by the ketocyclazocine class compounds. [2]The so-called sigma ligands such as SKF 10, 047 and nalorphine all have some degree of antagonist activity as well as being hallucinogenic.

stimuli as stress, nicotine or nerve stimulation and that the enkephalins can inhibit the release of catecholamines. S. Udenfriend and his colleagues have also shown that the adrenal chromaffin granules contain putative enkephalin-like precursor proteins and a heterogeneous mixture of enkephalin related peptides including met-enkephalin, leu-enkephalin and the heptapeptide tyr-gly-gly-phe-met-arg-phe[12]. It seems perfectly feasible that these peptides may have an intra-adrenal action on both the cortex and medulla and an extra-adrenal hormonal action on other peripheral tissues. The adrenal opioid peptides are hypotensive and it is possible that they may also be involved in peripheral vasodilator mechanisms.

Gastro-intestinal roles

It is interesting to speculate that one of the most widely used actions of opiates is for the relief of dysentery and diarrhoea and yet we are still far from understanding the mechanism of action. We are similarly ignorant of the gastro-intestinal functions of the endogenous opioid peptides. It is quite possible that pituitary and/or adrenal circulating opioid peptides may act on the gut and its associated structures, however the gastrointestinal tract is itself richly endowed with stores of the opioid peptides. There is in fact a greater total of enkepha-

lins in the gut than in the brain on an organ basis. Methionine-enkephalin and leucine-enkephalin are the predominant opioid peptides to be found in the gut and are mainly concentrated in nervous structures[13], although they are also present in secretory cells of the stomach, jejunum and pancreas. There are very much smaller amounts of β-endorphin-like peptides (less than 1% of enkephalin content) which appear to be localized in secretory cells.

The actions of opioids on the gut seem to be directed towards an inhibition of neuronal activity, inhibition of hormone release (including somatostatin and secretin) and inhibition of exocrine pancreatic secretion and of water transfer into the intestinal lumen. There is no evidence for a direct action on intestinal smooth muscle. The effects of opioid peptides on gastric secretion are controversial; morphine is known to augment oxyntic cell secretion but this is probably related to histamine release by the alkaloid since the enkephalins are apparently without effect[14].

The constipating effect of the opioids classically involves a decrease in gastric contractile activity and delayed emptying and inhibition of intestinal propulsive activity and of secretions. The overall effect is increased intestinal tone, a lengthened transit time and increased fluid readsorption. These effects may involve the inhibition of acetylcholine release and that of the non-cholinergic, non-adrenergic, intrinsic inhibitory transmitter. Inhibition of fluid transfer is probably associated with opiate receptor mediated inhibition of prostaglandin stimulated adenyl cyclase activity.

The role of endogenous opioids in peristalsis has received particular attention. Narcotic antagonist treatment of the isolated guinea-pig, rabbit, rat and cat ileum Trendelenburg preparation causes a stereospecific increase in peristaltic activity[15]. The overall effect of naloxone is to decrease the interval between peristaltic waves and to prolong the bursts of peristaltic activity. These effects are accompanied by a slight increase in smooth muscle tone. Opioids and opioid peptides have opposite effects. These results are consistent with a neuronal role for the enkephalins, in modulating the intrinsic reflex arc that controls peristalsis.

Concluding remarks

Research into the physiological roles of the opioid peptides has been facilitated by the availability of relatively specific antagonists. Yet it is apparent that still more specific antagonists are required. At the present time there is suggestive but not compelling evidence for both humoral and neurohumoral roles for the endogenous opioid peptides. Their wide distribution and overlapping actions and receptor specificities confuse any interpretation of the roles of the individual peptides. However many of these issues should be resolved if research continues at its present pace in this field. It is likely that the endogenous opioid peptides are quite ancient in phylogenetic terms, extending in distribution down to coelenterates. Their overall role has probably changed with evolution and it is my view that they act as modulators rather than as prime movers in physiological processes. It remains to be seen whether new therapeutic principles or applications arise from opioid peptide research but the present outlook is encouraging.

TABLE IV. Opioid peptides with possible peripheral activities

Y-G-G-F-M	Methionine-enkephalin
Y-G-G-F-L	Leucine-enkephalin
Y-G-G-F-M-K	Adrenal peptide
Y-G-G-F-M-R-F	Adrenal peptide
Y-G-G-F-L-R-R-I-R-P-K-K-G-Q-?	Dynorphin
Y-G-G-F-M-T-S-E-K-S-Q-T-P-L-V--T-L-F-K-N-A-H-K-K-G-Q	β-Endorphin

Code F = phenylalanine
K = lysine
N = asparagine
Q = glutamine
R = arginine
Y = tyrosine
All other letters refer to first letter of amino acid e.g.
G = glycine.

Reading list

1 Hughes, J., Kosterlitz, H. W. and Smith, T. (1977) Br. J. Pharmacol. 61, 639–647
2 Schulzberg, M., Lundberg, J. M., Hökfelt, T., Terenius, L., Brandt, J., Elde, R. P. and Goldstein, M. (1978) Neuroscience 3, 1169–1186
3 Schultzberg, M., Hökfelt, T., Terenius, L., Elfvin, L.-G., Lundberg, J. M., Brandt, J., Elde, R. P. and Goldstein, M. (1979) Neuroscience 4, 249–270
4 Robson, L. E. and Kosterlitz, H. W. (1979) Proc. R. Soc. Lond. (B) 205, 425–432
5 Wüster, M., Schulz, R. and Hertz, A. (1980) Life Sci. 14, 163–170
6 Hiller, J. M. and Simon, E. J. (1979) Eur. J. Pharmacol. 60, 389–390
7 Jean-Baptiste, E. and Rizack, M. A. (1980) Life Sci. 27, 135–142
8 Arrigo-Reina, R. and Ferri, S. (1980) Eur. J. Pharmacol. 64, 85–88
9 Goldstein, A., Tachibana, S., Lowney, L. I., Hunkapiller, M. and Hood, L. (1979) Proc. Natl. Acad. Sci. U.S.A. 76, 6666–6670
10 Kumakura, K., Guidotti, A., Yang, H.-Y. T., Saiani, L. and Costa, E. (1980) in Neural Peptides and Neuronal Communication (Costa, E. and Trabucchi, M., eds), Raven Press
11 Viveros, O. H., Diliberto, E. J., Hazum, E. and

Chang, K.-J. (1979) *Mol. Pharmacol.* 16, 1101–1108

12 Stern, A. S., Lewis, R. V., Kimura, S., Rossier, J. and Udenfriend, S. (1979) *Proc. Natl. Acad. Sci. U.S.A.* 76, 6680–6683

13 Schultzberg, M., Hökfelt, T., Nilsson, G., Terenius, L., Rehfield, J. F., Brown, M., Elde, R. P., Goldstein, M. and Said, S. (1980) *Neuroscience* 5, 689–744

14 Gascoigne, A. D., Hirst, B. H., Reed, J. D. and Shaw, B. (1980) *Br. J. Pharmacol.* 69, 527–534

15 Kromer, W., Pretzlaff, W. and Woinoff, R. (1980) *Life Sci.* 26, 1857–1865

Opiate receptors: some recent developments

E. J. Simon

Department of Psychiatry and Pharmacology, New York University Medical Center, 550 First Avenue, New York N.Y. 10016, U.S.A.

Two discoveries made in the first half of the 1970s sparked the enormous research activity that has characterized the opiate field in recent years. The first was the biochemical demonstration in 1973 of the existence of stereospecific opiate binding sites in the central nervous system of animals and man. These sites have been called opiate receptors and, for convenience, the author will follow this common usage. The reader should, however, be aware that the term receptor usually implies both a binding site and a transducing element that permits the binding of suitable ligands to trigger events that lead to the physiological response. Little is yet known about this transduction process or about the events set off by opiate binding, though much effort is going into bringing some light into this 'black box'. The excellent correlation between binding affinities of a large number of opioids and their pharmacological potencies provides evidence that the stereospecific sites are indeed the recognition and binding portion of opiate receptors.

The second important discovery was the demonstration, two years later, that the brain synthesizes molecules that possess morphine-like properties and are the putative endogenous ligands of the opiate receptors. Those endogenous opioids that have been well characterized to date turn out to be peptides and are known as enkephalins (pentapeptides) and endorphins (longer polypeptides).

This brief review will focus on some of the more recent advances that have been made in our knowledge of the opiate receptor. Other articles in *TIPS* will deal with the endogenous opioid peptides. I shall discuss some of the studies that lead us to believe that several classes of opiate receptors exist and I shall summarize progress that has been made in the solubilization of opiate receptors which are normally tightly bound to cell membranes. This is an important first step (one that proved difficult) towards the eventual purification of these receptors. For reviews of earlier work in this area see references 1 and 2.

Multiple opiate receptors

How many classes of opiate receptors are there? This is a question that has received a great deal of attention recently in a number of laboratories. It is an important question from both a theoretical and practical point of view. The question has been raised because of the many kinds of responses elicited by the opiate analgesics. These responses might be mediated via different types of receptors. There are several opioid peptides, each of which could exert its effects via its own receptor. Moreover, since most neurotransmitters so far examined act on multiple receptors, this could also be true for each of the opioid peptides, or at least those that represent neurotransmitters. From a theoretical viewpoint it is very important to know how many opiate receptors exist and what their functions are. The practical implications derive from the possibility that each sub-group of receptors may possess functional specificity. Thus, if a given sub-group is responsible for analgesia, while others mediate addiction and other undesirable side effects, it may become possible to synthesize molecules that have high affinity for the 'analgesic receptor', but low affinity for all others.

The hypothesis that multiple classes of opiate receptors exist was first examined by Martin and co-workers using classical pharmacological approaches in the chronic spinal dog[3,4]. It was found that the pharmacological spectra of different types of synthetic and natural opiates differed from

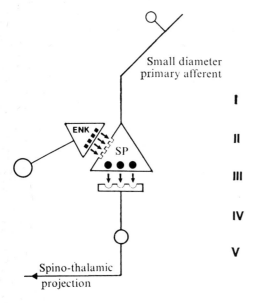

Fig. 1. Schematic representation of a possible mechanism for opiate-induced suppression of substance P (SP) release. SP is shown localized within the terminal of a small diameter afferent fiber which forms an excitatory axodendritic synapse with the process of a spinal cord neurone originating in lamina IV or V and projecting rostrally. A local enkephalin-containing inhibitory interneurone (ENK), confined to laminae II and III forms a presynaptic contact on the terminal of the primary afferent. Opiate receptor sites are depicted presynaptically. Roman numerals on the right refer to the laminae of Rexed. From Jessel, T. M. and Iversen. L. L. (1977) Nature (London) 268, 549–551. Reprinted by permission.

each other. Moreover, when the dogs were addicted to a given opiate, there were analogs that were unable to substitute for the chronically given drug in the suppression of withdrawal symptoms. Based on these studies Martin concluded that there are at least three types of opiate receptors in the CNS of the dog. He named them for the prototype drugs used: μ for morphine, κ for ketocyclazocine and σ for SKF-10047.

Stimulated by these studies, Hans Kosterlitz and his collaborators[5] were the first to attempt to determine whether similar sub-classes of receptors could be recognized through *in vitro* studies. They first examined potencies of opioids in the bioassay systems, used so effectively by this group, and found that enkephalins exhibit greater potency in the mouse vas deferens than in the guinea pig ileum, whereas the opposite was true for classical opioid alkaloids, such as morphine, levorphanol, etc. These results suggested that different populations of opiate receptors predominate in each of the two *in vitro* systems.

When cross competition for stereo-specific binding sites was studied in guinea pig brain membrane preparations, a result was obtained that seemed to complement the findings in the *in vitro* bioassays. Enkephalins proved to be much better in competing for binding with labeled enkephalins than with alkaloids. Alkaloids in turn, were more effective competitors against their own kind. Enkephalinamides (free carboxyl terminal blocked) proved to be about equally good competitors against both types of ligands. Data of this kind obtained in our own laboratory are shown in Table I and are very similar to the results published earlier by Lord *et al*[5].

Another approach used in our laboratory lends further support to the two-receptor concept. Jeffrey Smith and I studied the inactivation of opioid binding by the irreversible SH reagent, N-ethylmaleimide (NEM) and the ability of receptor ligands to protect[6]. We found that opiate alkaloids afforded better protection of alkaloid binding ([3H]naloxone or naltrexone) than of peptide binding, whereas enkephalins and their stable analogs were much more effective in protecting the binding of labeled peptides ([3H]Dala2-leu-enkephalin) against inactivation by NEM. Enkephalinamide was approximately equipotent regardless of ligand used (Table II). Simultaneously Robson and Kosterlitz[7] obtained virtually identical results using the irreversible inhibitor phenoxybenzamine. These results strongly support the existence of two classes of receptors and are, in my view, difficult to interpret in any other way.

If two or more classes of opiate receptors exist it should be possible to show a differential distribution of these receptors in regions of the CNS. Some success has indeed been achieved. In collaboration with Kenneth Bonnet[8] we carried out cross competition and saturation experiments in membrane preparations derived from various rat brain regions. Results in many regions resembled those in whole brain (Table I). The most striking difference was obtained in the thalamus, where naloxone was equally potent in competing with [3H]leu enkephalin as with labeled naloxone, whereas in whole brain and in most brain regions naloxone is 6–10 times more potent in competition against itself. The data suggest that the thalamus contains mainly receptors of the morphine-preferring or μ type. This result was supported by the saturation curves which gave straight line Scatchard plots with appropriate dissociation constants for μ receptor binding. An enrichment in δ receptors, though less clear cut, was observed in the amygdala. Similar results have been reported from P. Cuatrecasas' laboratory[9].

The existence of Martin's κ and σ receptors in rat brain has been suggested by pharmacological experiments but has not yet been confirmed by direct binding experiments. Several laboratories, including our own[10], have reported that the binding of [3H]ethylketocyclazocine, perhaps the best κ prototypic ligand, is essentially indistinguishable in properties, affinities and distribution from the binding of typical μ ligands, such as naloxone or morphine in rat brain.

Evidence for a specific binding site for phencyclidine (PCP) has recently been reported by Vincent et al. and Zukin and Zukin. The Zukins have also found that the opiate cyclazocine, a putative σ ligand, competes with PCP. They have suggested that the PCP site may, in fact, be identical to the σ opiate receptor. This would be an

TABLE I. Competition for opiate receptor binding

Competing compound	Labeled ligand	
	[3H]naloxone	[3H]leu-enkephalin
	nM	nM
Naloxone	2	12
Morphine	13	150
Etorphine	1	2.5
Met-enkephalin	30	3
Met-enkephalin amide	14	7
Leu-enkephalin	80	8
D-Ala2-leu-enkephalin	19	3
D-Ala2-met-enkephalin	20	0.5
D-Ala2-met-enkephalinamide	2	3

The numbers listed are IC50 values. All binding experiments were carried out at 0° for 150 min in the presence and absence of a large excess of unlabeled ligand corresponding to the tritiated ligand used. [3H]naloxone and [3H]leu-enkephalin were used at 1 nM. The results are the average of at least three closely similar experiments.

extremely interesting result, if supported by further experiments, since the hallucinogenic properties of cyclazocine and SKF-10047 resemble those of PCP.

Mention should be made of a different approach to receptor classification. Agu and Candace Pert have utilized sensitivity and lack of sensitivity to inhibition by GTP (or its non-hydrolyzable analog) to define two classes of opiate receptors (Type 1, inhibited and Type 2, not inhibited). Type 1 appears to correspond closely to 'μ' receptors and type 2 to 'δ' receptors.

M. Herkenham and C. B. Pert[11] have been able to map these receptors in a number of brain regions using autoradiography and have reported differences in their distribution.

One difficulty with this classification concerns the opiate receptors present in neuroblastoma x glioma hybrid cell cultures, which have been shown to be almost entirely of the enkephalin-preferring (δ) type. These receptors are coupled to adenylate cyclase and opiate as well as enkephalin binding are inhibited by GTP. They therefore seem to fit neither type 1 nor type 2.

After all that has been said, it should be pointed out that existence of two or more opioid receptors has not yet been proven. Even if they exist, the differences seem to be subtle ones. Thus, the inhibition by NEM and phenoxybenzamine is identical for alkaloid and enkephalin binding and differences are seen only when ligand protection is studied. Similarly, rates of enzymatic inactivation by either proteolytic enzymes or phospholipase A_2 are independent of ligand used to measure residual binding. Reactivation of phospholipase-inactivated opiate receptors by bovine serum albumin is also equally effective regardless of receptor ligand examined[12].

Equally perplexing are the results of Pasternak[13] who has examined the effect of several methods of inhibition on the Scatchard plot of putative μ, κ, σ and δ ligand binding. All inhibitors studied, including the putative affinity label naloxazone, synthesized by Pasternak, seem to diminish or wipe out the 'high affinity portion' of the curvilinear Scatchard, regardless of ligand used. This suggests identity of high affinity opiate-binding sites for these prototype ligands, seemingly at odds with the data discussed above.

The observed heterogeneity of opiate receptors could be the result of the existence of different receptor molecules or due to the presence of different 'prosthetic' groups on basically identical 'aporeceptors'. The possibility that identical receptors can exist in slightly altered conformations under different environmental conditions must also be considered. Much work is required to permit final conclusions to be reached about the nature of heterogeneity and about the number of classes of opiate receptors.

Receptor solubilization

One of the important goals in this research is to be able to purify the receptor

TABLE II. Ligand concentrations yielding half-maximal protection from inactivation by NEM

| Protecting agent | Protection of binding of | |
	[³H]naltrexone	D-Ala²-D-Leu⁵-[³H-]enkephalin
Morphine	100	800
D-Ala²-met-enkephalin	300	10
Naltrexone	3	60
D-Ala²-leu-enkephalin	400	60
D-Ala²-Met-enkephalinamide	70	40

The values given represent the concentration (nM) of protecting agents required to produce 50% of the maximal protection achieved for each labeled ligand. (From Ref. 6).

to homogeneity. This will permit elucidation of the chemical composition and subunit structure of the receptor and will make possible accurate kinetic and thermodynamic experiments, like those performed with pure enzymes. This would also permit the production of anti-receptor antibodies. The use of such antibodies as well as membrane reconstitution experiments would be extremely useful in aiding our understanding of opiate receptor function.

The opiate receptor is tightly bound to cell membranes and resisted for several years attempts in many laboratories to solubilize it. One of the reasons for these disappointing results was the extreme sensitivity of opiate receptors to low concentrations of detergents, even the non-ionic variety, which have been used to solubilize other receptors, such as those for acetylcholine and insulin.

A few years ago our laboratory[14] succeeded in solubilizing a macromolecular complex of [³H]etorphine using the non-ionic detergent, BRIJ 36T. This complex had properties consistent with its being a drug–receptor complex and had a molecular weight of about 400,000. This was the first step, but the solubilization of an opiate receptor that maintained its ability to bind opioid ligands in solution eluded us until recently. Dr Suzanne Zukin[15] repeated our experiments with identical results. She then demonstrated that a macromolecular complex of Dala²-leu-enkephalin with very similar properties could also be solubilized by BRIJ 36T. Using the cross linking reagent dimethylsuberimidate she succeeded in covalently linking some of the enkephalin analog to what appears to be an opiate receptor. This covalent complex of molecular weight 400,000 gave a band of apparent subunits of molecular weight 35,000 on SDS-polyacrylamide gel electrophoresis.

In a recent communication we reported that, in collaboration with Urs Ruegg[15], we have finally been able to solubilize active opiate receptors from toad brain by the use of digitonin. This method works very well for toad and bullfrog brain but has not been successful with mammalian brains. The idea of trying to solubilize a non-mammalian receptor occurred to Ruegg and was based on the experience of others with the β-adrenergic receptor which has been successfully solubilized in active form only from non-mammalian sources, such as turkey and frog erythrocytes. The properties of the soluble receptor strongly resemble those of the membrane-bound receptors suggesting that detergent treatment and removal of the receptor from its membrane environment does not significantly damage the binding site for opiates, at least by the criteria so far examined. To date, only a modest purification of about 6-fold has been obtained, but attempts to achieve a high degree of purity are in progress.

Two other laboratories have now reported the solubilization of active opiate receptors. Using a new detergent, nicknamed CHAPS, synthesized by L. M. Hjelmeland, Werner Klee[17] and his collaborators have succeeded in solubilizing opiate receptors from neuroblastoma x glioma hybrid cells. Some success has also been reported with receptors from rat brain.

J. Bidlack and L. Abood[18] have reported the solubilization of opiate receptors from rat brain. This was done with Triton X-100 which was subsequently removed with Biobeads SM2, since this detergent inhibits opiate binding even at minute concentrations. This emphasizes the importance of experimental conditions, since our laboratory and Klee's laboratory (private communication) tried this method unsuccessfully several years ago.

It is now clear that at least the binding site of the opiate receptor can survive solubilization. Rapid progress in purification can now be anticipated. I should temper this optimistic statement by pointing out that opiate receptors represent a minute portion of total brain protein and that the task of purification may be further complicated by the probable existence of two or more classes of opiate receptors.

Concluding remarks

Writing a brief essay about a small portion of such an active field is a frustrating experience. Even the two areas I covered have been reviewed in a less than exhaustive manner and with a bias towards our own studies. Much could have been said about the exciting studies on the sequelae of opiate binding, especially the studies on adenylate cyclase regulation in neuroblastoma x glioma cell cultures as well as the evidence for the involvement of this enzyme in the CNS. The use of a promising organotypic tissue culture system consisting of fetal mouse spinal cord with attached dorsal root ganglia by Stanley Crain in collaboration with our laboratory has already yielded some interesting data on the endogenous opioid system not easily obtained in intact animals or in established cell lines. Much more can be said, and hopefully will be said in *TIPS*, about the astonishingly rapid progress in the areas of biosynthesis, metabolism, distribution and function of the endogenous opioid peptides. For example, the molecular biology of the gene of the 31K precursor of β endorphin has already been worked out in considerable detail, giving an idea of the rate of progress in this research field.

Current Contents recently listed the 100 most active areas of research in the life sciences and the number of papers published in each in 1978. 'Opiate receptors and endorphins' won hands-down with 947 papers published (second place 645). In spite of all this activity we know surprisingly little about the physiological role of the endogenous opioid system. There is evidence that suggests that some of the opioid peptides act at the synapse as neuromodulators, i.e. their release modulates the rate of release of neurotransmitters. A model involving enkephalin and substance P is shown in Fig. 1.

A role for the opioid system in the endogenous (and exogenous) modulation of pain is supported by considerable, though largely still circumstantial, evidence. An important function of the system in opiate addiction is considered highly probable. However, in spite of the fact that much of this progress came from laboratories interested in the biochemical basis of opiate addiction, there is as yet little evidence for such a role.

The literature is replete with reports that suggest participation of opiate receptors and their ligands in alcohol abuse, overeating, several other normal and abnormal behaviors and a number of mental diseases, especially schizophrenia and depression. Much of this work is based entirely on the ability of opiate antagonists like naloxone to reverse the behavior or alter the pathology. Future research will tell which of these results represent a causal involvement of the endogenous opioid system.

We have every reason to be optimistic that work in this area and in related areas of other neuropeptides and their receptors (areas that have been largely spawned by the findings in the opiate field) will lead us to a better understanding of the biochemical basis of brain function and behavior.

Reading list

1 Simon, E. J. and Hiller, J. M. (1978) *Ann. Rev. Pharmacol.* 18, 371–394
2 Beaumont, A. and Hughes, J. (1979) *Ann. Rev. Pharmacol.* 19, 245–267
3 Martin, W. R., Eades, C. G., Thompson, J. A., Huppler, R. E. and Gilbert, P. E. (1976) *J. Pharmacol. Exp. Ther.* 197, 517–532
4 Gilbert, P. E. and Martin, W. R. (1976) *J. Pharmacol. Exp. Ther.* 198, 66–82
5 Lord, J. A. H., Waterfield, A. A., Hughes, J. and Kosterlitz, H. W. (1977) *Nature (London)* 267, 495–499
6 Smith, J. R. and Simon, E. J. (1980) *Proc. Natl. Acad. Sci. U.S.A.* 77, 281–284
7 Robson, L. E. and Kosterlitz, H. W. (1979) *Proc. R. Soc. Lond. B.* 205, 425–432

8 Simon E. J., Bonnet, A., Crain, S. M., Groth, J., Hiller, J. M. and Smith, J. R. (1980) in *Neural Peptides and Neuronal Communication* (Costa, E. and Trabucchi, M., eds), pp. 335–346, Raven Press, New York

9 Chang, K. J., Cooper, B. R., Hazum, E. and Cuatrecasas, P. (1979) *Mol. Pharmacol.* 16, 91–104

10 Hiller, J. M. and Simon, E. J. (1980) *J. Pharmacol. Exp. Ther.* 214, 516–519

11 Herkenham, M. and Pert, C. B. (1980) *Proc. Natl. Acad. Sci. U.S.A.* 77, 5532–5536

12 Lin, H. K. and Simon, E. J. (1978) *Nature (London)* 271, 383–384

13 Pasternak, G. W. (1980) *Proc. Natl. Acad. Sci. U.S.A.* 77, 3691–3694

14 Simon, E. J., Hiller, J. M. and Edelman, I. (1975) *Science* 190, 389–390

15 Zukin, R. S. and Kream, R. M. (1979) *Proc. Natl. Acad. Sci. U.S.A.* 76, 1593–1597

16 Ruegg, U. T., Hiller, J. M. and Simon, E. J. (1980) *Eur. J. Pharmacol.* 64, 367–368

17 Simonds, W. F., Koski, G., Streaty, R. A., Hjelmeland, L. M. and Klee, W. A. (1980) *Proc. Natl. Acad. Sci. U.S.A.* 77, 4623–4627

18 Bidlack, J. M. and Abood, L. G. (1980) *Life Sci.* 27, 331–340

Histamine H₁ receptors in the CNS

J. M. Young

Department of Pharmacology, University of Cambridge, Hills Road, Cambridge, CB2 2QD, U.K.

Despite the explosion of activity in the receptor labelling field in recent years, the histamine H_1 receptor has been surprisingly neglected. The synthesis of a suitable ^3H-ligand has filled this gap and provided the means to obtain information on the properties and distribution of histamine H_1 receptors in the CNS.

Writing in the foreword to Volume XVIII/I of the Handbook of Experimental Pharmacology, Histamine and Antihistamines, in 1966 Sir Henry Dale noted that "more than a thousand pages . . . now (seem) to be required for the review of all the separate aspects of what has so far been discovered and discussed about this one, simple, natural base, histamine". Of these 1000 pages the chapter 'Histamine in the brain' occupied a mere eight and comparison of its content with the admirable review of histaminergic mechanisms in the CNS by Schwartz[9] in 1977 makes clear how much progress has been made in the last two decades, even though the functional role or roles the amine may play remain largely uncertain.

The actions of histamine in the CNS appear to be mediated by both H_1 and H_2 receptors, but whereas the evidence for H_2 receptors in 1977 was good, the presence of H_1 receptors was much less certain. The object of this brief review is to present recent evidence, largely obtained from receptor binding studies, for the properties and distribution of histamine H_1 receptors in the CNS.

Development of [³H]mepyramine as an H₁ ligand

Our own involvement with central histamine H_1 receptors developed from work with [³H]propylbenzilylcholine mustard on the complexity of muscarinic agonist binding to intestinal muscle strips. The actions of acetylcholine and histamine on the guinea-pig ileum have many similarities and it seemed reasonable to suppose that if the complex curves for muscarinic agonists represented some essential feature of receptor function, then histaminergic agonists should show the same behaviour. It was primarily to test this that we set out to synthesize a ^{3}H-ligand for histamine H_1 receptors. It seems strange that with so many antihistamines described in the literature, many with apparently eminently suitable properties, the H_1 receptor should have been one of the last of the 'classic' receptors to be subjected to binding studies.

The choice of ^{3}H-ligand was not difficult. The 'typical' H_1 antagonist mepyramine (pyrilamine) offered a number of advantages – good receptor selectivity, a dissociation constant of the order of magnitude desired, i.e. *circa* 1 nM, and a convenient synthetic route to the ^{3}H-compound. The last was achieved by bromination of mepyramine to the mono-bromo derivative, which underwent a smooth reduction with tritium gas back to [³H]mepyramine[6]. The same synthetic route can be used for other 2-aminopyridinyl-antihistamines, such as tripelennamine and methapyrilene, but mepyramine had another particular advantage in that the bromo-derivative was distinctly weaker than mepyramine itself as an H_1 antagonist, so that the concentration, and hence chemical purity, of the ^3H-ligand, initially determined by u.v. absorbance, could be checked by bioassay on the guinea-pig ileum.

Initial binding studies with [^3H]-mepyramine were carried out on homogenates of the longitudinal muscle from guinea-pig small intestine[5], the tissue on which the most organ-bath measurements had been made. The good agreement between affinity constants determined by the two methods left little doubt that [^3H]mepyramine binds selectively to H_1 receptors and opened the way for the use of this ligand by ourselves[2-4] and by Snyder and his co-workers[1,12] to investigate histamine H_1 receptors in the CNS.

H_1 receptors in guinea-pig brain

Guinea-pig brain contains appreciable amounts, mean value 227 ± 52 pmol/g protein, of [^3H]mepyramine binding sensitive to inhibition by promethazine, triprolidine or other H_1 antagonists. The evidence that the antagonist-sensitive binding represents binding to histamine H_1 receptors is strong. The promethazine-sensitive binding is saturable and gives a K_d for [^3H]mepyramine in reasonable agreement with that obtained from inhibition of the contractile response of guinea-pig ileum to histamine. All the potent antagonists tested inhibit the binding of 1 nM [^3H]mepyramine to the same extent as promethazine and the affinity constants derived from these curves are in good agreement with those from organ-bath studies. Finally the potency ratio for the stereoisomers of chlorpheniramine as

inhibitors of [^3H]mepyramine binding is closely similar to that reported on the ileum. The identity of the properties of peripheral and central H_1 receptors is striking.

The regional distribution of H_1 receptors in guinea-pig brain is markedly uneven[1,2], with the highest amounts being found in the cerebellum, moderate amounts in the hypothalamus, hippocampus and thalamus and rather less in the basal ganglia. Some of the H_1 receptors may well be associated with blood vessels, but it seems unlikely that this is their sole location. There is, for instance, no evidence of any gross differences in blood flow between cerebellum and caudate nucleus, yet the numbers of H_1 receptors differ by a factor of 6.

The regional distribution of H_1 receptors does not parallel that of histamine, histidine decarboxylase, histamine-N-methyltransferase or numbers of mast cells. However, in view of the complexity of the histamine system in brain it is probably unreasonable to expect a simple correlation with any one of these. Histamine is stored in both neuronal and non-neuronal compartments. The former is the only one associated with the specific histidine decarboxylase, whilst the latter, which has a much slower turnover probably represents storage in mast cells[9]. This split distribution makes it likely that histamine has at least two distinct functional roles, possibly more, and only certain of these might be mediated through H_1 receptors. Certainly there is good evidence for the presence of H_2 receptors from the inhibition by metiamide or cimetidine of depressant effects of microiontophoretically applied histamine on the firing of cortical and hippocampal cells and from the inhibition by the H_2 antagonists of the histamine-stimulated adenylate cyclase activity in homogenates from several brain regions.

Correlation of the density of either H_1

or H_2 receptors with amounts of histamine-N-methyltransferase seems unlikely, since the enzyme is fairly evenly distributed in brain tissue[11] and may not be associated with the sites of primary action of the amine, in analogy with the lack of parallelism in the cholinergic system of the distribution of acetylcholinesterase and that of choline acetyltransferase.

Evidence for H_1 receptor functions in the CNS

The properties of the promethazine-sensitive [³H]mepyramine binding provide strong evidence for the presence of sites with H_1 character, but give no indication what response – if any – they might mediate. The evidence for H_1 functions in the brain of the guinea-pig or any other species, with properties very similar to those of peripheral H_1 receptors is limited. Often H_1-mediated actions have been inferred from the use of H_1 antagonists at concentrations at which non-specific effects cannot be ruled out. Histamine induced hypertension, hypothermia, emesis and water intake have been proposed on the grounds of antagonist selectivity to be H_1 actions and are discussed in the earlier review[9].

The best quantitative evidence for H_1 receptors seems to be that of Schwartz and his co-workers[7] who have shown that activation of the histamine-sensitive adenylate cyclase in slices of guinea-pig hippocampus appears to be mediated by both H_1 and H_2 receptors, although the dissociation constants determined for triprolidine, mepyramine, diphenhydramine and, most notably, promethazine, were somewhat lower than the values obtained from antagonism of the contractile response of guinea-pig ileum to histamine. However Chasin et al. failed to find any evidence for a histamine stimulated cyclase in slices of guinea-pig cerebellum[9].

Comparatively little has been published on the effects of histamine on cyclic GMP formation in brain, although the presence of an H_1-sensitive histamine-stimulated guanylate cyclase has been reported by Richelson[8] in mouse neuroblastoma cells and by Study and Greengard[10] in bovine superior cervical ganglion. The affinities of mepyramine and diphenhydramine for these systems were reasonably similar to the values determined from guinea-pig ileum, although, as discussed below, comparison of constants determined in different species must be made with caution.

Species variation

Discussion of [³H]mepyramine binding has thus far been restricted to data obtained on the guinea-pig, since this is the species for which the best case can be made that the binding in brain homogenates is to histamine H_1 receptors. However, it has become apparent that there are distinct species variations in the pattern and characteristics of [³H]mepyramine binding in mammalian brain. The differences were first apparent in comparing the regional distribution of [³H]-mepyramine in the brain of the rat and the guinea-pig and have been extended to calf and human brain by Snyder's group. The apparent number of H_1 receptors in whole brain homogenates (frontal cortex only in the case of human brain) was similar in all four species[1], but whereas the cerebellum contained the highest density of receptors in the guinea-pig, in both human and rat brain, the cerebellum was among the regions with a relatively low H_1 receptor content. The notable feature of H_1 receptor distribution in human brain is the preponderance in cerebral cortical areas[1].

The species variation is not confined to differences in regional distribution of [³H]mepyramine binding. There are also species differences in the structure of the binding site, as indicated by the variation of the dissociation constants of H_1 antagonists. Thus Snyder's group have reported[1] constants (nM) for mepyramine of 0.5 (guinea-pig), 1.0 (human), 3.0 (mouse), 3.9 (rabbit) and 4.0 (rat), while our own measurements[4] have given values of 0.8 (guinea-pig) and 9.0 (rat). The values for (+)-chlorpheniramine, triprolidine and chlorpromazine show a similar species variation, but those for certain other antagonists, e.g. promethazine, diphenhydramine and chlorcyclizine, do not. It is notable, however, that the apparent affinity of histamine as an inhibitor of [³H]mepyramine binding does not vary greatly between species, so that although the H_1 receptors may not be identical, there need be no difference in sensitivity to the natural agonist.

Differences between the guinea-pig and the rabbit in the response of aortic strips to histamine have been recorded in the literature and the K_d values reported for mepyramine, 1.2 nM (guinea-pig) and 3.9 nM (rabbit), are closely similar to those from binding data. For other species, the similar pattern of antagonist potency in inhibiting [³H]mepyramine binding suggests that the sites to which the ³H-label binds are H_1 receptors, but it would be valuable to have confirmatory data from tissue responses – if indeed central and peripheral H_1 receptors do show similar properties in these species. It is clearly also necessary to be sure that the ³H-ligand is binding to a single set of sites in a particular species. In the rat there is some evidence for low-affinity promethazine-sensitive [³H]mepyramine binding sites in both brain and intestinal smooth muscle[4]. Indeed, in the latter tissue, which in our animals gave no indication of a mepyramine-sensitive contractile response to histamine, the greater part of the promethazine-sensitive binding may be to low affinity sites.

H_1 receptors and clinical response

The problem of species variation must be borne in mind when using [³H]-mepyramine binding as a test system with which to examine the effects of drugs on H_1 receptors, but with due caution the simplicity of the method makes it attractive for use in assessing the central anti-H_1 actions of therapeutic agents. The potency of tricyclic antidepressants, including iprindole, as inhibitors of [³H]-mepyramine binding has been noted[12] and the question asked whether H_1 receptor blockade could in part be responsible for their antidepressant action. Certain of the antischizophrenic drugs, notably chlorpromazine and clozapine[1,3], are also potent H_1 blockers and even though there is no correlation between H_1 block and antischizophrenic activity, the point has been made[3] that with some of the drugs H_1 block will occur at clinical doses and could contribute to the overall clinical response. Similarly it should be possible to assess whether the well-known sedative and performance impairing effect of most of the antihistamines commonly used against hay-fever is directly related to central anti-H_1 potency.

In conclusion

It will be only too apparent from this brief review that knowledge of central histamine H_1 receptors is still limited and that although binding studies with [³H]-mepyramine should prove a useful tool with which to investigate these receptors, so far they have raised more questions than they have answered. There are in addition complexities which may prove significant, but which in the space available it has not been possible to touch on. The binding of [³H]mepyramine in guinea-pig brain or intestine seems to fit well to a simple hyperbola and the curves for the inhibition of [³H]mepyramine binding by most of the potent antagonists we tested had the shape expected for a simple equilibrium with a single set of sites. However, the curve for mepyramine itself did not – seemingly, to use the words of Sir Henry again, to be "another of those paradoxes, which the study of histamine has already appeared to present in an unending series". And perhaps it should be added, because this is where we came in, that apparently the curve for histamine is not simple either.

Reading list

1. Chang, R. S. L., Tran, V. T. and Snyder, S. H. (1979) *J. Neurochem.* 32, 1653–1663.
2. Hill, S. J., Emson, P. C. and Young, J. M. (1978) *J. Neurochem.* 31, 997–1004.
3. Hill, S. J. and Young, J. M. (1978) *Eur. J. Pharmac.* 52, 397–399.
4. Hill, S. J. and Young, J. M. (1980) *Br. J. Pharmac.* 68, 687–696.
5. Hill, S. J., Young, J. M. and Marrian, D. H. (1977) *Nature (London)* 270, 361–363.
6. Marrian, D. H., Hill, S. J., Sanders, J. K. M. and Young, J. M. (1978) *J. Pharm. Pharmac.* 30, 660–662.
7. Palacios, J. M., Garbarg, M., Barbin, G. and Schwartz, J. C. (1978) *Mol. Pharmac.* 14, 971–982.
8. Richelson, E. (1978) *Nature (London)* 274, 176–177.
9. Schwartz, J.-C. (1977) *Annu. Rev. Pharmac. Toxicol.* 17, 325–339.
10. Study, R. E. and Greengard, P. (1978) *J. Pharmac. Exp. Ther.* 207, 767–778.
11. Taylor, K. M. (1975) In: L. L. Iversen, S. D. Iversen and S. H. Snyder (eds), *Handbook of Psychopharmacology,* Vol. 3, Plenum Press, New York and London, pp. 327–379.
12. Tran, V. T., Chang, R. S. L. and Snyder, S. H. (1978) *Proc. Nat. Acad. Sci. U.S.A.* 75 6290–6294.

GABA receptors

S. J. Enna

Departments of Pharmacology and of Neurobiology and Anatomy, University of Texas Medical School at Houston, P.O. Box 20708, Houston, Texas 77025, U.S.A.

γ-Aminobutyric acid (GABA) appears to be an important neurotransmitter in the mammalian central nervous system (CNS)[1-4]. Support for this proposition is provided by the fact that GABA and the enzymes responsible for its synthesis and degradation are heterogeneously distributed throughout the neuroaxis and, like other neurotransmitter candidates, GABA is concentrated in a distinct synaptosomal population and there exists a high affinity transport and calcium-dependent release mechanism for this substance. Furthermore, immunocytochemical and autoradiographic studies have identified morphologically distinct GABA-containing neurons in brain and spinal cord.

Perhaps the most compelling evidence to support a transmitter role for this amino acid is provided by electrophysiological studies demonstrating that GABA causes a conductance change in neurons leading to a hyperpolarization that is identical to that produced by the endogenous inhibitory transmitter[5]. The specificity of this receptor action was established by the finding that the convulsants picrotoxin and bicuculline are able to selectively block this effect of GABA[6]. Further proof for the existence of neuronal GABA receptors has been provided by the identification and characterization of these sites using ligand binding assays *in vitro*[7].

These discoveries, coupled with the fact that alterations in the GABAergic system may account for the symptoms of a variety of neuropsychiatric disorders[1,8], have spawned a great deal of research aimed at defining the biochemical, pharmacological and functional properties of GABA receptors. In the present communication the results of these studies will be briefly summarized and discussed. Since it is possible to highlight only some of the major advances in this area, readers desiring more detailed information are urged to consult any of a number of other, more comprehensive, reviews and monographs[1-4].

Electrophysiological, biochemical and pharmacological properties

GABA is considered an inhibitory neurotransmitter even though this substance can induce both a hyperpolarizing and depolarizing response. Both of these actions are thought to be the consequence of a GABA receptor mediated alteration in chloride conductance. Thus, the GABA-induced hyperpolarization that is typically observed in cerebral cortical cells is brought on by an increase in the flow of chloride ion into the postsynaptic element resulting in an outward flow of current which, in turn, moves the membrane potential towards the chloride equilibrium potential, converting it to the classic inhibitory postsynaptic potential (IPSP).

The depolarizing response to GABA, which predominates in spinal cord cells, is also inhibitory since it represents a depolarization of presynaptic terminals, leading to a decrease in the amount of transmitter released (presynaptic inhibition or primary afferent depolarization). In this case it is thought that the GABA receptor mediated increase in chloride conductance results in an outward flow of chloride, presumably because the intracellular chloride concentration in these presynaptic terminals exceeds the extracellular concentration of this ion. It is possible, however, that presynaptic inhibition may also be the result of an accumulation of extracellular potassium ions which are released as a consequence of the discharge of neighboring afferent fibers. With respect to biochemical studies, ligand binding assays have revealed that the synaptic GABA receptor is a complex structure composed of a variety of independent, but interrelated, sites. Thus, binding studies have revealed that the agonists [³H]GABA and [³H]muscimol label the recognition site of this receptor[7,9]. The binding of the radiolabelled antagonist [³H]bicuculline is also to the recognition site, though it displays somewhat different biological and pharmacological characteristics[10], suggesting that there may be a conformational difference between the agonist and antagonist states for this receptor. In contrast, [³H]dihydropicrotoxinin, another GABA receptor antagonist, does not attach to the recognition site but rather appears to label the chloride channel[11].

Other studies have revealed the presence of a substance(s), on or near the recognition site, which regulates the affinity of this site. The removal of this substance or modulator, either by extensive washing or treatment with Triton X-100, unmasks a group of higher affinity GABA receptor recognition sites[7]. The importance of this modulator in GABA receptor function is suggested by the discovery that benzodiazepines may activate the GABAergic system by displacing this substance, thereby increasing the sensitivity of GABA receptors[12].

Further evidence linking the GABA and benzodiazepine receptor sites is provided by the finding that activation of GABA receptors induces an increase in benzodiazepine receptor binding[13]. This enhancement in benzodiazepine binding is due primarily to a GABA receptor mediated increase in the affinity of benzodiazepine receptor sites.

Thus, there are several loci on GABA receptors which, when activated or inhibited, can alter GABA receptor function. With regard to the recognition site, a number of compounds have been identified as direct acting agonists or antagonists for this portion of the receptor. Included among the agonists are the systemically active compounds muscimol (3 - hydroxy - 5 - aminomethylisoxazole), THIP (4,5,6,7-tetrahydroisoxazolo (5,4-c) pyridone-3-ol), SL-76002 [a(chloro-4-phenyl) fluro-5 hydroxy-2 benzilidene-amino-4H-butyramide] and kojic amine (2 - aminomethyl - 5 - hydroxy - 4H - pyran - 4-one)[14]. Other agonists are 3-aminopropane sulfonic acid, imidazole-4-acetic acid, trans-4-aminocrotonic acid and 4-aminotetrolic acid[15]. However, because of poor lipid solubility, none of this latter group appears to be active following systemic administration.

A number of phthalideisoquinolines appear to be active as recognition site antagonists. Included in this group are bicuculline, corlumine and hydrastine[16].

With regard to the ionophore portion of the receptor, no compounds have yet been unequivocally identified as specific agonists for this site. However, ligand binding data have suggested that certain depressant barbiturates such as mephobarbital, metharbital and pentobarbital may act in this manner[11]. In contrast, both electrophysiological and biochemical data indicate that picrotoxin, tutin, tetramethylenedisulphotetramine, some silatranes, a number of bicyclophosphates, t-butylbicyclocarboxylates and some convulsant barbiturates may all act by blocking the GABA receptor chloride ionophore.

Finally, GABA receptor activity may also be modified by agents which alter the concentration of the proposed recognition site modulator. At present, however, only the benzodiazepines appear to modify GABA receptor activity in this way, though it is conceivable that other drugs, hormones, or disease states may alter the function of GABA receptors by influencing the concentration of this substance.

GABA receptor subtypes

It has been suspected for some time that, as in other neurotransmitter systems, there exist pharmacologically and functionally distinct types of GABA receptors. Although electrophysiological studies have suggested the existence of bicuculline-insensitive GABA receptors, it has been difficult to characterize this site further because of a lack of appropriate pharmacological tools. Due to this limitation, cautious investigators have defined GABA receptors strictly on the basis of sensitivity to bicuculline and/or picrotoxin. Thus, to date, only bicuculline-sensitive GABA receptors have undergone extensive investigation.

Recent studies have hinted at the possibility that there may be subtypes of bicuculline-sensitive GABA receptor sites. For example, pretreatment of brain membranes with Triton X-100 reveals two kinetically distinct populations of GABA receptors, each with a different regional distribution[12]. While these sites appear to be pharmacologically indistinguishable, there are data to suggest that only the lower affinity site is linked to the benzodiazepine receptor[17], indicating that these sites, in addition to being kinetically unique, may also be functionally distinct. Moreover, recent immunocytochemical and radiohistochemical studies have also suggested that the benzodiazepine binding site may be associated with only a limited number of GABA receptors[18,19].

It is also noteworthy that while GABA and muscimol are capable of activating benzodiazepine receptor binding, other electrophysiologically active GABA receptor agonists, such as THIP, are inactive in this regard[17]. One explanation for this finding is that while GABA and muscimol are pure receptor agonists, THIP may be only a partial agonist for this site. However, it is also conceivable that the inability of THIP to activate benzodiazepine receptor binding may indicate that THIP is a selective agonist for GABA receptors not linked to the benzodiazepine system. If so, then THIP or related agents may be good candidates for labelling a subpopulation of GABA receptors. In general, however, the issue of GABA receptor subtypes is unresolved and will remain so until compounds can be found which will activate or inhibit separate populations of receptors. Such a discovery would be of immense value in pointing the way for the development of more selective, and perhaps less toxic, GABAmimetic agents.

GABA receptor function

Since GABA receptor pharmacology is still in its infancy there is little precise information about the functional consequences of GABA receptor activation, although indications are that this receptor regulates a variety of behavioral, biochemical and physiological activities[3]. For example, GABA receptor agonists appear to be anticonvulsants and anxiolytics. Furthermore, receptor stimulation decreases aggressive behavior in rat, and hypothalamic GABA receptors modulate feeding behavior and satiety.

With regard to biochemical responses, no evidence has yet been supplied to indicate that these receptors are linked directly to specific enzymes on postsynaptic membranes, though it seems reasonable to assume that there is some coupling to an enzyme system which helps maintain the membrane pump for the transport of chloride ion. However, studies *in vivo* have suggested that the cerebellar cyclic GMP and pituitary cyclic AMP systems may be indirectly influenced by GABA receptor activation since inhibition of the GABAergic system leads to significant increases in the tissue content of these nucleotides. Activation of GABA receptors also appears to modify dopamine turnover and dopamine, acetylcholine and perhaps even GABA release in various brain regions. GABA receptors also influence hormonal activity by modulating the release of prolactin, growth hormone, thyrotropin and gonadotrophin.

GABA receptor activation has also been reported to modify heart rate and blood pressure, causing a dose-dependent reduction in both parameters. These responses appear to be centrally mediated, at least in part by GABA receptors located on brain stem neurons in the vicinity of the nucleus ambiguous.

Thus, GABA receptors regulate a myriad of biological processes. Interestingly, however, clinical studies have suggested that perhaps some GABAmimetics are more selective with regard to these actions than others. For example, at the doses used clinically, neither SL-76002 nor THIP has yet been found to have a significant effect on heart rate or blood pressure (B. Scatton and P. Krogsgaard-Larsen, personal communication). This preliminary finding may be taken as further evidence for the existence of functionally distinct types of GABA receptors, or it may indicate that other GABA receptor actions are capable of counteracting the cardiovascular responses in man.

Summary and conclusion

Bicuculline-sensitive GABA receptors are found in all areas of the mammalian CNS[20]. Activation of these receptors induces a shift in membrane permeability to chloride causing a depolarizing (presynaptic inhibition) or hyperpolarizing (postsynaptic inhibition) response, depending upon the relative intracellular concentration of this ion. Because of the large number of these receptor sites, and their ubiquitous distribution, it appears likely that there may be pharmacologically and functionally distinct GABA receptors, although conclusive data are still lacking in this regard.

Biochemical and electrophysiological studies have indicated that the GABA receptor is a complex of separate, but interrelated, sites. Thus, there are two kinetically distinct types of recognition site, and these recognition sites may have two conformations, one favoring agonist and the other antagonist attachment. Furthermore, the ionophore portion of the receptor, while physically and pharmacologically distinct, is regulated by the recognition site, and some GABA receptor recognition sites are functionally linked to benzodiazepine receptor recognition sites such that activation of either leads to an enhancement in the receptor affinity for the other. Finally, located on or near the GABA receptor recognition site is an endogenous modulator which regulates the affinity of this site. Because of this complexity, GABA receptor function can be pharmacologically manipulated in a variety of ways. Accordingly, the GABA receptor should not be envisaged as a singular entity, but rather as a symphonious family of sites, making overall receptor action dependent upon the status of each individual component.

Acknowledgements

Preparation of this manuscript was facilitated by United States Public Health Service grant and NS-13803, and Research Career Development Award (NS-00335) and by grants from Mead Johnson and Co. and Merck, Sharp and Dohme.

Reading list

1 Enna, S. J. (1979) *Ann. Rep. Med. Chem.* 14, 41–50
2 Roberts, E., Chase, T. N. and Tower, D. B. (1976) *GABA in Nervous System Function* Raven Press, New York

3 Krogsgaard-Larsen, P., Scheel-Kruger, J. and Kofod, H. (1979) *GABA-Neurotransmitters* Academic Press, New York

4 Pepeu, G., Kuhar, M. J. and Enna, S. J. (1980) *Receptors for Neurotransmitters and Peptide Hormones* Raven Press, New York

5 Krnjevic, K. and Schwartz, S. (1967) *Exp. Brain Res.* 3, 320–36

6 Curtis, D. R., Duggan, A., Felix, D. and Johnston, G. A. R. (1971) *Brain Res.* 32, 69–96

7 Enna, S. J. and Snyder, S. H. (1977) *Mol. Pharmacol.* 13, 442–453

8 Enna, S. J. (1980) *Can. J. Neurol. Sci.* 7, 257–259

9 Enna, S. J., Beaumont, K. and Yamamura, H. I. (1978) in *Amino Acids as Chemical Transmitters* (Fonnum, F., ed.), pp. 487–492, Plenum Press, New York

10 Mohler, H. and Okada, T. (1978) in *Amino Acids as Chemical Transmitters* (Fonnum, F., ed.), pp. 493–498, Plenum Press, New York

11 Olsen, R. W., Ticku, M. K., Greenlee, D. and Van Ness, P. (1979) in *GABA-Neurotransmitters* (Krogsgaard-Larsen, P., Scheel-Kruger, J. and Kofod, H., eds), pp. 165–178, Academic Press, New York

12 Costa, E. (1979) *Trends Pharmacol. Sci.* 1, 41–44

13 Tallman, J. R., Thomas, J. W. and Gallager, D. W. (1979) *Nature (London)* 274, 383–385

14 Enna, S. J. and Maggi, A. (1979) *Life Sci.* 24, 1727–1738

15 Johnston, G. A. R., Allan, R. D., Kennedy, S. M. E. and Twitchin, B. (1979) in *GABA-Neurotransmitters* (Krogsgaard-Larsen, P., Scheel-Kruger, J. and Kofod, H., eds), pp. 149–164, Academic Press, New York

16 Johnston, G. A. R. (1976) in *GABA in Nervous System Function* (Roberts, E., Chase, T. N. and Tower, D. B., eds), pp. 395–411, Raven Press, New York

17 Braestrup, C., Nielsen, M., Krogsgaard-Larsen, P. and Falch, E. (1980) in *Receptors for Neurotransmitters and Peptide Hormones* (Pepeu, G., Kuhar, M. J. and Enna, S. J., eds), pp. 301–312, Raven Press, New York

18 Palacios, J. M., Unnerstall, J. R., Young, W. S. and Kuhar, M. J. (1980) in *GABA and Benzodiazepine Receptors* (Costa, E., DiChiara, G. and Gessa, G., eds), pp. 53–60, Raven Press, New York

19 Mohler, H., Richards, G. and Wu, J. C. (1980) in *GABA and Benzodiazepine Receptors* (Costa, E., DiChiara, G. and Gessa, G., eds), pp. 139–146, Raven Press, New York

20 Enna, S. J. (1979) in *GABA-Biochemistry and CNS Functions* (Mandel, P. and DeFeudis, F., eds), pp. 323–337, Plenum Press, New York

The role of gamma-aminobutyric acid in the action of 1,4-benzodiazepines

E. Costa

Laboratory of Preclinical Pharmacology, National Institute of Mental Health, St Elizabeths Hospital, Washington, D.C. 20032, U.S.A.

Although many derivatives of 1,4-benzo-diazepines have been among the best selling drugs for the last 15 years, the synaptic mechanisms whereby these drugs exert their anticonvulsant and anxiolytic effects began to be elucidated only recently. The present report concerns this elucidation and extensively deals with the contributions made by my collaborators, Drs Guidotti, Mao, Biggio, Suria, Toffano, Baraldi, Leon and Schwartz to the present understanding of the mode of action of 1,4–benzodiazepines. I have attempted to contrast our work with a description of simultaneous and preceding research developments on the mechanisms of action of these fascinating centrally acting drugs in order to justify our inferences and conclusions.

A. The origin of 1,4–benzodiazepines

The story of 1,4–benzodiazepines began in Krakow during the mid-thirties when Dr Leo H. Sternbach considered the 4,5–benzo–[hept]–1,2,6–oxdiazepines as prospective starting materials for a new series of dyes[1], but this work failed to yield important results and was abandoned. After a silence of several years, it was again Dr Sternbach, now at Hoffman La Roche Labs in Nutley, N.J., who proposed the heptoxdiazepines as a starting material to produce a new series of tranquillizers[2]. Chlordiazepoxide was the first product of this series to reach pharmacological screening.

It was Dr L. Randall[3] who by selecting an appropriate battery of pharmacological tests could differentiate the pharmacological profile of chlordiazepoxide from that of meprobamate, chlorpromazine and barbiturates. Dr Randall concluded that chlordiazepoxide was a centrally acting compound with a unique pharmacological profile and predicted its potential significance in therapy.

This prediction was verified, as shown by the enthusiasm the new drug generated after only a few years of clinical trials. In a very short time, chlordiazepoxide became the number one prescription item all over the world as a drug of choice for the treatment of anxiety, neuroses and psychosomatic disturbances of various kinds. During the last 10 years, chlordiazepoxide has no longer been the most frequently prescribed drug because diazepam, another derivative of 1,4–benzodiazepine, has taken its place. The pharmacological profile of diazepam is quite similar to that of chlordiazepoxide but diazepam is more potent. In elucidating the mechanism of 1,4–benzodiazepines, their capacity to antagonize seizures elicited by blockers of gamma-aminobutyric acid (GABA) receptors or metabolism became of value[4,5]. However, this facilitation of GABAergic mechanisms elicited by 1,4–benzodiazepines is not used to screen active congeners of these drugs and still today, it is felt that in rats, the best laboratory tests to help predict the therapeutic efficacy of 1,4–benzodiazepines in patients, is their ability to restore the behavioral output that is depressed by response contingent punishment or conditioned fear[5]. Since this action of 1,4–benzodiazepines is blocked by picrotoxin[6] and is facilitated by other manipulations which facilitate GABAergic transmission, it appeared reasonable to us[4] to suggest that the synaptic mechanisms which participate in the mediation of the anticonflict (anxiolytic?) action of 1,4–benzodiazepines are also GABAergic. However, there is no unanimity of consensus on this inference. Some behavioralists presently believe in an as yet unspecified unique mechanism that underlies the action of 1,4–benzodiazepines on punished behavior. This belief is strengthened by the finding that with muscimol, a powerful GABA receptor agonist, it is difficult to obtain in rats a dose-dependent stimulation of the behavior that is depressed by response contingent punishment or conditioned fear. However, in contrast, muscimol can mimic a number of actions of diazepam including the antagonism of the convulsions elicited by isoniazid, picrotoxin and bicuculline and the induction of a selective decrease in the cyclic GMP content of cerebellum in rats[4]. Undoubtedly some aspects of the pharmacological profile of 1,4–benzodiazepines and muscimol are different, but these differences could reflect a specific heterogeneity of the GABA receptors[7] rather than the participation of another synaptic mechanism. It will be discussed later in this report how the 1,4–benzodiazepines selectively modify the affinity of one of the two GABA receptors present in brain[7]. According to this view, 1,4–benzodiazepines facilitate GABAergic transmission only in those receptors where the affinity for GABA is regulated by a receptor modulator protein[8]; in contrast, muscimol, a direct agonist that can bind directly to GABA receptors acts on both receptors simultaneously. This difference in the mode of action of muscimol and 1,4–benzodiazepines will be clarified by highlighting the current understanding of the mechanisms that are operative in GABAergic transmission[7-10].

CHLORDIAZEPOXIDE   DIAZEPAM

B. GABA as a transmitter

Like other neurotransmitters, GABA also causes specific changes in the properties of the postsynaptic membrane. To discuss these changes, let us take as a reference the fluid mosaic model of the cell membrane. According to this widely accepted model, the postsynaptic receptors, including GABA receptors are conceived as a supramolecular complex of two or more protein molecules embedded in the lipid bilayer of the membrane of the postsynaptic cell. The various components of the GABA receptor are assumed to cluster and interact with each other when GABA binds to the recognition site. It is currently believed that second messenger mechanisms (cyclases) or ionophores are coupled to postsynaptic receptors to help them to bring intracellularly the messages that trans-synaptically reach the extracellular sites on the membrane. The cyclases are membrane enzymes which respond to the occupancy of the receptors by increasing the rates of cyclic nucleotide formation inside the cell. The ionophores are also integral protein of membranes which exist in two different conformational states; their transition is controlled allosterically by mechanisms triggered by the occupation of postsynaptic receptors. One ionophore conformation facilitates the influx of specific ions. When GABA occupies the postsynaptic receptors the Cl^- ionophore can function and the Cl^- fluxes through the membrane according to the concentration gradient. When the extracellular fluid contains more Cl^- than the intracellular fluid of the cell that is innervated by GABAergic synapses, the trans-synaptic release of GABA increases the Cl^- fluxes and hyperpolarizes the cell membrane. Consequently, the intensity of this cell response to incoming excitatory (depolarizing) stimuli is reduced or abolished. The kinetic of the binding of GABA to the specific recognition sites in the membrane of postsynaptic neurons suggests that there is more than one type of GABA recognition site; however, we know very little about the functional role of these biochemically heterogeneous binding sites for GABA. This heterogeneity is revealed by the different kinetics of GABA binding to freshly prepared or to frozen, thawed and Triton X-100 treated brain membranes[7,8]. This treatment increases the total number of GABA binding sites; these newly appearing GABA receptors have a dissociation constant smaller than that of the GABA receptors that are present in freshly prepared membranes; hence the treatment with Triton X-100 unveils a new population of GABA receptors. This detergent appears to act by displacing a receptor modulator protein with a molecular weight of about 1.5×10^4 dalton which by an allosteric mechanism can modify the conformation of the GABA recognition site, and thereby prevent the expression of the high affinity binding sites for GABA[7,8].

This peripheral membrane protein is a very powerful inhibitor of Ca^{2+} dependent and independent protein kinase. When the synaptic junctions are thoroughly washed from this protein, their phosphorylation increases by 20-fold or more. Since this protein changes the affinity of recognition sites for GABA by modulating the phosphorylation of membrane proteins it has been termed GABA-modulin. The view that phosphorylation is involved comes from experiments showing that if the synaptic junctions are previously phosphorylated the GABA-modulin fails to change the affinity of the recognition sites for GABA.

The data of Table I show that the membranes of various brain areas have a different density of GABA receptors and that this difference is due mainly to a different density of GABA receptors with low affinity ($GABA_1$). In cerebellum the density of these $GABA_1$ receptors is almost 10-fold greater than that of the substantia nigra; but in nigra and striatum the GABA receptors with a high affinity ($GABA_2$ receptors) are 57 and 41 per cent of the total GABA binding, respectively. Since the density of $GABA_2$ receptors is more or less constant in all the brain areas studied while that of $GABA_1$ receptors changes from structure to structure, it follows that in the hypothalamus the percentage of the total GABA receptors represented by $GABA_2$ receptors is greater than in cortex whereas this percentage is very small in cerebellum. The $GABA_2$ receptors are those in which a specific protein via an allosteric mechanism regulates the number and affinity of the GABA binding sites. We will discuss later experiments showing that diazepam can displace this modulator protein and, therefore, facilitate the GABAergic transmission by increasing the number of high affinity $GABA_2$ receptors. Not all direct GABA receptor agonists have a marked (10-fold) difference in their affinity for $GABA_1$ and $GABA_2$ receptors. In the case of muscimol, for instance, the difference in the affinity for $GABA_1$ and $GABA_2$ receptors is much less.

C. GABA-modulin regulates the affinity of GABA to the recognition sites of $GABA_2$ receptors

GABA-modulin is the endogenous thermostable membrane protein which is located in the junctional membranes of several types of brain cells and can interact with GABA receptors by allosterically inhibiting their capacity to bind GABA[8]. This regulatory function is mediated via membrane protein phosphorylation; the protein that is phosphorylated has not yet been isolated. The high affinity binding of radioactive naloxone, QNB, dihydroalprenolol and spiroperidol is not modified by the GABA-modulin[7,8]. However, this receptor modulator protein cannot be viewed as an exclusive regulator of GABA receptors, because it can modulate the benzodiazepine binding to synaptic membranes[7-9]. The characteristics whereby this protein binds to cell membranes suggest that it may be a peripheral membrane protein because it can be released easily by washing the membranes with buffer. We have purified the protein by about 1000-fold[7] and found that the properties to inhibit cyclic AMP-dependent and independent protein kinases and to modify $GABA_2$ receptors are purified by identical extents. Though the changes in affinity of recognition sites for GABA were measured in enriched junctional membrane preparations from brain homogenates (unpublished results) it was difficult to exclude the possibility that the GABA-modulin was bound to the membrane of neurons that do not possess GABAergic innervation and that as an artifact of homogenization this protein

TABLE I. The density of regular and low affinity GABA receptors in synaptic membranes of various brain areas

Brain area	GABA receptor density		
	Total	Low affinity $GABA_1$	Regular affinity $GABA_2$
Substantia nigra	1.75	0.75	1.00
Striatum	1.93	1.10	0.83
Hypothalamus	4.15	3.10	0.95
Cortex	5.65	4.90	0.75
Cerebellum	7.40	6.50	0.90

Units to describe receptor density (B_{max} = pmol/mg membrane protein), binding studies were performed at 0° and in Na^+ free medium with frozen, thawed brain membranes washed with Triton X-100. (Modified from Guidotti et al.[9].)

was brought to interact with GABA receptors. Such an artifact can be excluded by virtue of recent studies conducted by Drs Guidotti, Baraldi and Schwartz with neuroblastoma cell lines. A fresh membrane preparation of this cell line contains only one population of GABA receptors which like the GABA receptors present in freshly prepared brain membranes have a low affinity for GABA (K_d 200 nM). When these membranes are frozen, thawed and then treated with Triton X-100 the modulator protein that inhibits the high affinity binding of GABA (GABA$_2$ receptor) can be removed and a second receptor population (GABA$_2$) with high affinity sites (K_d 20 nM) can be expressed. One possibility presently under investigation is that GABA$_1$ and GABA$_2$ receptors of this cell line modulate the Cl$^-$ ionophore. Preliminary results indicate that the activation of GABA receptor causes an increase in the inward flux of Cl$^-$. Moreover a cooperative interaction between GABA and 1,4–benzodiazepine receptors was found to extend also to the regulation of Cl$^-$ fluxes. Since it is possible to sensitize to GABA the recognition site of the GABA$_2$ receptor by removing the endogenous protein regulator, one wonders whether there is an endogenous agonist continuously performing this function. From the kinetic studies conducted it is possible to infer that GABA may act as the endogenous agonist that removes GABA-modulin and thereby by facilitating phosphorylation unveils high affinity GABA$_2$ receptors which when activated open up the Cl$^-$ channels. Also 1,4–benzodiazepines act as specific competitive antagonists of GABA-modulin, and thereby facilitate GABAergic transmission by unveiling high affinity recognition sites of GABA$_2$ receptors. Since 1,4– benzodiazepines also facilitate protein phosphorylation in junction membranes they must be viewed as a special kind of specific indirect agonist of GABA$_2$ receptors and, therefore, different from muscimol, which is an agonist of low affinity GABA$_1$ and unveils high affinity GABA$_2$ receptors.

D. Strategies used to detect the mode of action of 1,4–benzodiazepines

Theoretically, drugs can affect brain function by either changing the velocity whereby stimuli are decrementlessly conducted in axons or by interfering with specific types of synaptic transmission. Since the regulation of conduction is presumably similar in various types of neurons, the action of drugs that act on conduction exhibits a limited degree of specificity in pharmacological action. In contrast, since specific mechanisms are operative in regulating different synapses, specific actions can be expected from drugs acting at synapses. Here drugs can act either at presynaptic or at postsynaptic sites. Presynaptically, drugs can interfere with transmitter synthesis, its storage in vesicles, the axonal transport of vesicles, the release or reuptake of transmitters in axon terminals. Postsynaptically, drugs can block the activation of the transmitter, they can mimick or antagonize the action of the transmitter, they can facilitate or reduce the response of the receptor to the transmitter by modifying mechanisms that regulate the affinity of postsynaptic receptors for the transmitter. Several lines of electrophysiological, biochemical, and pharmacological investigations suggest that the benzodiazepines facilitate GABAergic transmission[4,5,10]. By measuring the GABA turnover in brain areas which contain GABAergic synapses one can distinguish whether a drug that facilitates GABAergic transmission acts presynaptically or postsynaptically. If the drug under study increases GABAergic function by releasing GABA it will increase GABA turnover, in contrast, if the drug facilitates GABAergic transmission by a postsynaptic mechanism the turnover of GABA should not increase. When this criterion was applied to the study of 1,4–benzodiazepines, it was found that the turnover rate of GABA was not increased by these drugs and therefore, we concluded that the site of action of 1,4–benzodiazepines on GABAergic transmission was located postsynaptically[10].

E. Reciprocal interactions between brain receptors for GABA and benzodiazepines

Since 1,4–benzodiazepines fail to mimic the actions of GABA when they are applied to receptors after depletion of GABA[10] but they potentiate the responses to GABA when this transmitter is released by nerve impulses[4,5] it was concluded that the 1,4–benzodiazepines modify the susceptibility of GABA receptors by acting at specific regulatory sites of GABA receptors. When it was found that 1,4–benzodiazepines bind with a high affinity to brain membranes[11] it was implied that they occupy a receptor which is normally occupied by a transmitter or by an endogenous modulator[12]. The presence of GABA-modulin located on GABA receptors[8] invites speculation that the 1,4–benzodiazepine high affinity binding site is located in coincidence with a site that regulates $GABA_2$ receptors[9] via GABA-modulin. Neuroblastoma cells perfused with 1,4–benzodiazepines release a great amount of GABA-modulin. Moreover the addition of 1,4–benzodiazepines or muscimol to junctional membrane increases the Ca^{2+}-dependent phosphorylation of junctional membrane protein. We have shown that GABA-modulin decreases the affinity of the $GABA_2$ recognition sites for GABA by an allosteric mechanism and that 1,4– benzodiazepines displace the endogenous modulator protein competitively and thereby increase the affinity of the $GABA_2$ receptors[7–10]. Consistent with these data is the hypothesis that the saturable high affinity binding site for the 1,4–benzodiazepines[11] is different from that for GABA. Although any known transmitter, including GABA, fails to compete for the high affinity binding sites for the 1,4–benzodiazepines, the occupancy of the GABA receptor facilitates 1,4–benzodiazepine binding and the endogenous protein modulator displaces competitively the 1,4–benzodiazepines from their binding sites. In several brain areas, the density of high affinity binding sites for the 1,4–benzodiazepines decreases and increases with that of the GABA receptors[9]. However, it has been reported that when the GABA receptors are lesioned, the number of GABA and 1,4–benzodiazepine receptors does not decrease in parallel. Such a discrepancy would be expected if the 1,4–benzodiazepine receptors were located in proximity of the $GABA_2$ receptors, only.

The picture that emerges from experiments with crude synaptic membrane preparation obtained from brain is verified by experiments with a cell line of neuroblastoma (NB_{2a}) that contains in its membranes $GABA_1$ and $GABA_2$ receptors, the protein modulator and the high affinity binding sites for 1,4–benzodiazepines. Though precise computation cannot be made, it appears that in these membranes the content of GABA-modulin is greater than expected, if this receptor modulator were exclusively located in the 1,4–benzodiazepine receptors. We know that the endogenous protein inhibits the phosphorylation of proteins by protein kinase and we can infer that probably this protein regulates phosphorylation in several cell sites including the $GABA_2$ receptor sites. GABA-modulin competes with the high affinity binding of 1,4–benzodiazepines also, in this neuroblastoma cell line; moreover, also in these cells, like in brain, when GABA receptors are stimulated by GABA receptor agonists, the high affinity binding of 1,4–benzodiazepines is facilitated (M. Baraldi, A. Guidotti, J. Schwartz and E. Costa, unpublished).

This work on $NB_{2\alpha}$ membranes allows us to infer that although 1,4-benzodiazepine and $GABA_2$ receptors are two separate molecular entities, they are probably part of a single functional unit. In fact, the reciprocal changes in the affinity of GABA and 1,4-benzodiazepine receptors when one of the two receptors is occupied by the specific agonist suggest that the receptors for 1,4-benzodiazepines are part of a supramolecular functional entity, which includes GABA receptors, the GABA-modulin, the protein kinase and the ionophore. This unit operates according to the concept that receptors are supramolecular entities made up by a cluster of mobile proteins floating in the fluid mosaic of the postsynaptic membranes. Consistent with the latter hypothesis is the isolation of the GABA-modulin which modifies allosterically GABA binding to $GABA_2$ receptors and interacts competitively with the 1,4-benzodiazepines' specific binding[7-9]. Since a number of observations suggest that this apparent competitive interaction cannot be accounted for as a binding of 1,4-benzodiazepines to the GABA-modulin[9] we are proposing that by interacting with the binding of the GABA-modulin that inhibits the affinity of the recognition sites of the $GABA_2$ receptor, the 1,4-benzodiazepines regulate GABAergic transmission. In fact, we are presently investigating whether a GABA recognition site phosphorylation is operative in the allosteric regulation of GABA affinity for $GABA_2$ receptors; we have reason to believe that GABA itself is an agonist that by binding to the low affinity receptors releases GABA-modulin and by phosphorylation of a receptor protein unveils the high affinity GABA recognition sites which, when activated open the Cl^- channels.

F. Conclusions

The evidence discussed in this report suggests that the interaction between the 1,4-benzodiazepines and the allosteric protein modulator of $GABA_2$ receptors, is relevant to explain the facilitation of GABAergic transmission by 1,4-benzodiazepines[4,5]. Since the rank order of several 1,4-benzodiazepines to compete with the protein modulator is similar to the rank order of their potency in relieving anxiety[9], one can propose that the anxiolytic action of 1,4-benzodiazepines is mediated by their facilitation of GABA transmission at the $GABA_2$ receptors. It is possible that anxiety expresses a defect in the function of $GABA_2$ receptors due to an abundance of the receptor modulator protein, but this possibility must still be demonstrated. Certainly, 1,4-benzodiazepines are now not only a successful drug but a research tool to investigate GABAergic transmission and to study mechanisms of behavioral regulation, by GABA-modulin and membrane protein phosphorylation.

Reading list

1. Dziewonski, F. and Sternbach, L. H. (1933) *Bull. Int. Acad. Pol. Class Sci. Med. Nat. Ser. A*, 416.
2. Sternbach, L. H. (1978) In: E. Jucker (eds), *Progress in Drug Research*, Vol. 22, Birkhauser Verlag, Basel and Stuttgart, pp. 224-266.
3. Randall, L. (1960) *Diseases of Nervous System*, Vol. 21, Suppl. 2, p. 7.
4. Guidotti, A. (1978). In: M. A. Lipton, A. Di Mascio, K. F. Killam (eds), *Psychopharmacology: A Generation of Progress*, Raven Press, New York, pp. 1349-1358.
5. Haefely, W. E. (1978) In: M. A. Lipton, A. Di Mascio and K. F. Killam (eds), *Psychopharmacology: A Generation of Progress*, Raven Press, New York, pp. 1359-1374.
6. Stein, L., Wise, C. D. and Belluzzi, J. D. (1975) In: E. Costa and P. Greengard (eds), *Advances in Biochemical Psychopharmacology*, Vol. 14, Raven Press, New York, pp. 29-44.
7. Costa, E., Guidotti, A. and Toffano, G. (1978) *Br. J. Psychiatry* 133, 239-248.

8. Toffano, G., Guidotti, A. and Costa, E. (1978) *Proc. Nat. Acad. Sci. U.S.A.* 75, 4024–4028.
9. Guidotti, A., Toffano, G. and Costa, E. (1978) *Nature (London)* 275, 553–555.
10. Biggio, G., Brodie, B. B., Costa, E. and Guidotti, A. (1977) *Proc. Nat. Acad. Sci. U.S.A.* 74, 3592–3596.
11. Squires, R. F. and Braestrup, A. (1977) *Nature (London)* 269, 732–734.
12. Iversen, L. (1977) *Nature (London)* 266, 678–683.

Searching for endogenous benzodiazepine receptor ligands

Claus Braestrup and Mogens Nielsen

A/S Ferrosan, Sydmarken 5, DK-2860 Soeborg, Denmark and Sct. Hans Hospital, DK-4000 Roskilde, Denmark.

The discovery of benzodiazepine receptors in 1977 prompted an intensive search for endogenous ligands with benzodiazepine-like activity or the opposite. Derivatives of β-carboline 3-carboxylic acid are very potent and seem interesting. It is also interesting that benzodiazepine receptors and GABA receptors are tightly coupled at the molecular level.

Introduction

What is a receptor? The concept is not clearly defined, but it is something like: 'the particular biological macromolecule to which an agent binds with high structural selectivity and with the consequence that the characteristic biological activity of the agent is elicited'. We have to deal with several kinds of receptors, for example neurotransmitter receptors, hormone receptors, and drug receptors. It is clear that not all receptors are classified once and for all, because the classification depends on the biological role of the ligand, be it endogenous or not. Opiate receptors were classified at first as drug receptors with the use of high affinity binding techniques, and then eventually as neurotransmitter receptors in parallel with the gradual acceptance of the neurotransmitter role of some endorphins.

Address for correspondence:
Claus Braestrup, A/S Ferrosan, Sydmarken 5, DK-2860 Soeborg, Denmark.

Brain benzodiazepine receptors

1,4-Benzodiazepines are among the most widely used drugs. Approximately 15% of the adult population in the West ingests a benzodiazepine at least once a year to achieve anticonvulsant, anxiolytic, or hypnotic effects. Some benzodiazepines are very potent drugs. Clinical effects are achieved after as little as 30 μg kg^{-1} of triazolam (a triazolo-1,4-benzodiazepine), and 5 mg of diazepam to non-tolerant adults gives manifest responses.

The development of high affinity binding techniques provided the means for a direct approach to studying the mechanism of action of benzodiazepines at the molecular level. Tritium-labelled diazepam was added to crude rat brain membranes[11], and it appeared that a protein in these membranes avidly bound [³H]diazepam. Half maximal binding was obtained at only 3 nM for [³H]diazepam and less for [³H]flunitrazepam[1,11]. Scatchard analysis of data from saturation experiments yielded strictly straight lines, which indicated the presence of only a single class of binding sites and a limited number of binding sites[1]. Recent studies, however, have found indications of multiple classes of binding sites[12] and the presence of at least two distinct molecular species (51,000 and 55,000 daltons) with benzodiazepine receptor sites[7].

A binding site for a compound, however high its affinity or limited the number of sites it may have, should not be regarded as a 'receptor' unless some relation to relevant biological consequences can be established. Unfortunately there is no simple biological test for benzodiazepine action *in vitro*; peripheral tissues are virtually nonresponsive to benzodiazepines. That benzodiazepine binding sites are probably 'receptors' was established by showing that the affinity of several benzodiazepines for the binding sites *in vitro* (estimated as K_i values, see legend Fig. 2) was closely correlated to those pharmacological tests *in vivo* which predict clinical effects in man, namely inhibition of pentazol convulsions, cat muscle relaxant effect, Geller-anticonflict activity and others[1]. Another good argument in favour of the receptor status of [3H]diazepam binding sites is the intimate interaction with GABA receptors at the molecular level[13] (see below).

Benzodiazepine receptors are present only in brain tissue. Liver membranes, kidney membranes, and serum albumin bind benzodiazepines, but these sites are clearly different from the specific brain receptors[1]. No pharmacological or clinical effects have yet been attributed to peripheral benzodiazepine binding sites.

There is a regional receptor distribution in the brain of both rat and man. The highest receptor levels are present in cortical regions, a result which suggests a predominantly cortical action of benzodiazepines. Medium receptor concentration occurs in several nuclei, such as the corpus striatum and the thalamus. Low receptor densities occur in white matter areas, such as the corpus callosum and the medulla oblongata.

Autoradiographic techniques, using bound [3H]flunitrazepam, often reveal 'patchy' receptor appearances within a brain region. For example in lamina II, III and IV of the cervical spinal cord there is a restricted zone of high receptor concentration[8]. Gross regional studies did not indicate a preferential limbic occurrence of benzodiazepine receptors but autoradiography of amygdala, for example, pointed at certain nuclei with high receptor concentration.

Neuronal localization

It turned out to be quite tedious to show that benzodiazepine receptors are located on neurons. Autoradiographic techniques did not achieve the required resolution and even electron microscopic examination of autoradiograms of covalently bound [3H]flunitrazepam are inherently uncertain, although they indicate location at nerve terminals. Receptor antibodies are not available for immunohistological studies. However, neuronal localization was shown in biochemical experiments, using lesioning techniques. Neuronal degeneration in the corpus striatum, substantia nigra, and cerebellum of rats, induced by the neurotoxic agent kainic acid, caused extensive depletion (50–70%) of benzodiazepine receptors. The reductions occurred 10–30 days after lesions, depending on the brain region examined. In addition the level of benzodiazepine receptors was found to be reduced in the caudate nucleus and putamen of patients dying with Huntington's chorea, a disease with profound and well-described neuronal degenerations. Furthermore, a small (20%) decrease was demonstrable in the level of benzodiazepine receptors in the cerebellum of the mutant 'nervous' (nr/nr) mouse, which exhibits a selective degeneration of cerebellar Purkinje cells, but not in that of the weaver (wv/wv) mutant, which lacks cerebellar granule cells. These findings, coupled with the total lack of brain-specific benzodiazepine receptors on primary mouse astroglial cells, point to an almost exclusive neuronal localization on many, but not all kinds of brain neurons.

Benzodiazepines and GABA

GABA (γ-aminobutyric acid) mediates both pre- and post-synaptic inhibition in all regions of the CNS. GABA is accepted as the major inhibitory neurotransmitter of the brain. Electrophysiological studies show that benzodiazepines invariably increase the inhibitory effects of GABA on neurons[6]. In addition, several behavioural studies are compatible with the idea that some effects of benzodiazepines are mediated by GABA potentiating mechanisms, particularly the anticonvulsant effect.

That benzodiazepine receptors and GABA receptors are also tightly coupled at the molecular level is evident from experiments where GABAergic agents were added to thoroughly washed rat brain membranes. Washing was necessary to remove high amounts of endogenous GABA. GABA, muscimol, and nine other agonists doubled the binding of

[³H]diazepam to benzodiazepine receptors[2,13]. The affinity was increased rather than the maximal number of binding sites. The GABA-induced increase was stereoselectively inhibited by the GABA antagonist bicuculline which competes for the GABA recognition site, but not by picrotoxin which is believed to act mainly on the GABA-receptor-dependent chloride channel[13].

The GABA receptor recognition site involved in the *in vitro* increase in [³H]diazepam binding is probably very closely related, if not identical, to the GABA receptor which mediates physiological effects of GABA. Micromolar concentrations of GABA are effective, as expected from electrophysiological studies. The structural requirements reveal the expected rank order of agonists, and both the GABA antagonist, bicuculline, and agonists exhibit the expected stereoselectivity[2,13].

Surprisingly, a small group of well-characterized GABA mimetics was found to increase [³H]diazepam binding very

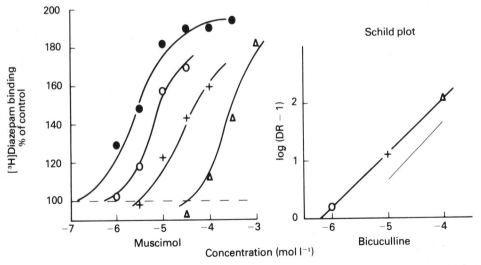

Fig. 1. Left. Dose–response curve for the effect of muscimol on [³H]diazepam binding to washed rat forebrain membranes. Various concentrations of (±)-bicuculline were added (●, no addition; ○, 10⁻⁶M; +, 10⁻⁵M; △, 10⁻⁴M). Right, Schild plot constructed from the data on the left. The dose ratio (DR) is the proportion between EC₅₀ values for muscimol with and without bicuculline added. The almost parallel displacement of dose–response curves by bicuculline (left) and the slope close to unity (right) indicates that muscimol and bicuculline compete for the same site in a first order interaction.

little or not at all [aminopropanesulphonic acid, isoguvacine, imidazoleacetic acid, β-guanidinopropionic acid, THIP (4,5,6,7-tetrahydroisoxazolo-[5,4-c]pyridin - 3 - ol), and piperidine 4-sulphonic acid]. Interestingly, however, all of the non-efficacious agonists were capable of antagonizing the effect of GABA and the agonist muscimol on benzodiazepine binding. The GABA receptor involved in the increase in [^{3}H]diazepam binding thus recognizes all the relevant chemical structures. The subgroup of GABA mimetics mentioned above behave as partial agonists or even as competitive antagonists with respect to at least one conformational change in the GABA receptor: the change affecting benzodiazepine receptors. They appear to be full agonists with respect to other conformational changes, i.e. for the change in GABA receptors that causes chloride channels to open in the spinal cord. 'Classical' pharmacological methods can be used on the GABA/benzodiazepine coupling because interaction with a receptor (the GABA receptor) causes an effect (increased affinity of the benzodiazepine receptor)[2]. We have exemplified this approach by a Schild-plot for bicuculline (Fig. 1).

In a recent issue of TIPS, Costa reviewed the evidence that GABA-modulin, a 15,000 dalton, heat stable protein, is responsible for a reciprocal benzodiazepine/GABA receptor interaction. GABA-modulin inhibits GABA receptors allosterically, and it is hypothesized that benzodiazepines enhance GABAergic transmission by displacing GABA-modulin, thereby disinhibiting GABA receptors[4].

The coupling between benzodiazepine receptors and GABA receptors may account for some of the pharmacological effects of benzodiazepines, notably the anticonvulsant effect, without inferring the presence of endogenous ligands. On the other hand it is unlikely that benzodia-

zepine receptors are exclusively coupled to the GABA system, because the major clinical feature of benzodiazepines, the antianxiety effect, has not been convincingly linked to the GABA system. Furthermore, there is a clear dissociation between the phylogenetic, ontogenetic and regional distributions of benzodiazepine receptors and GABA receptors. In addition, the level of benzodiazepine receptors is increased, while the GABA receptor level is decreased in the mouse mutant of the weaver type. These results show uncoupling, and leave us with neuronal benzodiazepine receptors without known physiological function. Obviously the presence of endogenous ligands is suspected.

Endogenous ligands

Some of the compounds shown in Fig. 2 have been purified from biological material by means of their affinity for benzodiazepine receptors. Hypoxanthine, inosine, and nicotinamide are present in brain and their pharmacological and electrophysiological effects are to some extent compatible with a benzodiazepine-like mode of action[9,10]. However, the affinity of hypoxanthine, inosine, and nicotinamide for benzodiazepine receptors is faint and it must be questioned whether they are available in adequate concentrations at the receptor sites in the brain.

β-Carboline 3-carboxylic acid as its ethyl ester (β-CCE) was isolated from 1800 liters of human urine[3]. From an affinity point of view, this compound is very attractive because it posseses even higher affinity for benzodiazepine receptors than diazepam. However, β-CCE is not present as such in any measurable concentration in brain or peripheral tissues in vivo. Furthermore, it is clear that manipulations during the isolation procedure, which at first were believed to cause hydrolysis of urinary conjugates, were esterifications. When extraction of β-CCE from brain tissue was attempted, the compound was only obtained when esterification with ethanol was included in the isolation procedure. Brain tissue contains high activities of carboxymethylase enzymes, which convert certain proteins into methyl esters using S-adenosylmethionine as the methyl donor. The methyl ester of β-CC is as active as the ethyl ester, but methylation of β-carboline 3-carboxylic acid with S-adenosylmethionine and brain enzymes has not been demonstrated. Derivatives of β-carboline 3-carboxylic acid other than the esters may play a role at brain benzodiazepine receptors.

β-CCE is quite specific for benzodiazepine receptors[3]. GABA receptors, dopamine receptors, acetylcholine receptors, and noradrenaline receptors, all identified by high-affinity binding techniques, are not affected by β-CCE. Interestingly, the only neurotransmitter receptors weakly affected (more than 100-fold less than the benzodiazepine receptor) were serotonin receptors, although this is not totally unexpected from inspection of the chemical structures. Serotonin is the only example amongst well-known putative neurotransmitters which seems to be involved in the anxiolytic action of benzodiazepines.

Most interestingly, β-CCE appears to exert a selective action on benzodiazepine receptors. [³H]Flunitrazepam can be displaced from cerebellar membranes with 5–7 times lower concentrations of β-CCE than those required to displace it from hippocampus, indicating the presence of regional variations in receptor characteristics. Multiplicity of benzodiazepine receptors, having different affinities for β-CCE may account for these regional differences.

β-Carbolines

β-Carbolines are not new to pharmacology and neurochemistry. The chemist Goebel isolated harmaline from the seeds of the Asian plant *Peganum harmala* in 1841. Harmaline, harmine (7-methoxy-1-methyl-β-carboline) and other substituted and/or partially hydrogenated

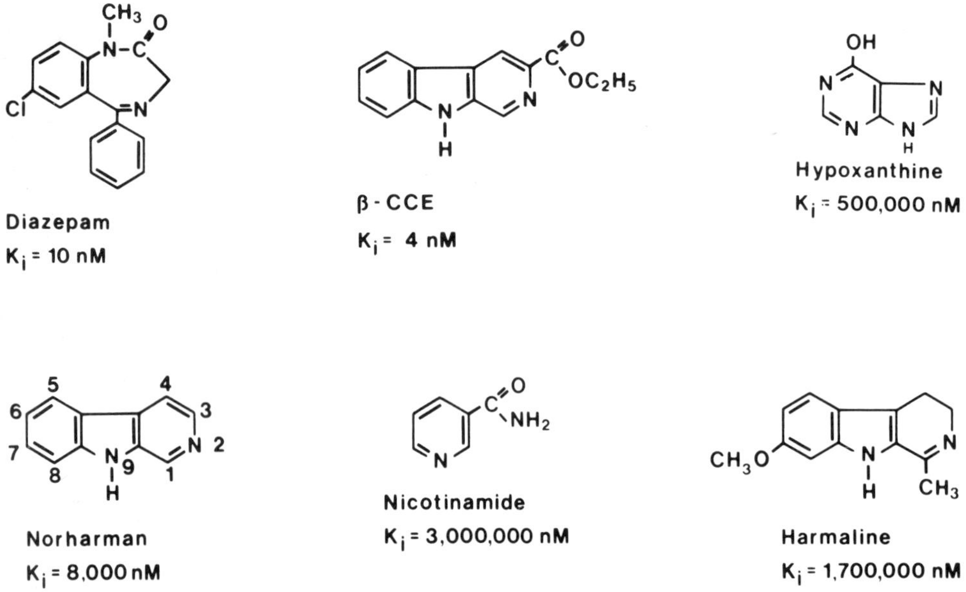

Diazepam
$K_i = 10\ nM$

β-CCE
$K_i = 4\ nM$

Hypoxanthine
$K_i = 500,000\ nM$

Norharman
$K_i = 8,000\ nM$

Nicotinamide
$K_i = 3,000,000\ nM$

Harmaline
$K_i = 1,700,000\ nM$

Fig. 2. Hypoxanthine, nicotinamide and β-carboline 3-carboxylic acid as the ethyl ester (γ-fraction, β-CCE) were isolated from biological sources by means of their affinity for benzodiazepine receptors. The K_i value was calculated from the IC_{50} value (Ref. 1) which is the concentration of the agent that inhibits 50% of specific [³H]diazepam binding to rat brain membranes. Low K_i value means high affinity.

β-carbolines were identified in the hallucinogenic extracts, called caapi or yage, prepared from the bark of *Banisteria caapi* by Amazonian tribes in South America. Naranjo has characterized the hallucinations induced by harmaline as superimposition of imaginary scenes on surrounding walls and ceilings. It is characteristic that the hallucination occurs simultaneously with undistorted perception. When the eyes are closed, the subject experiences long dreamlike sequences. Perception of music is not enhanced as after ingestion of LSD or mescaline. A methoxy-group in position 6 or 7 of the β-carboline is essential for hallucinogenic properties, but seems prohibitive for benzodiazepine receptor interaction (Fig. 2).

Harmaline and harmine are very potent, reversible MAO inhibitors of the A-selective type. It is not known whether MAO inhibition is related to the hallucinatory properties because harmaline exhibits several other pharmacological and neurochemical features such as stimulation of climbing fibres to the cerebellum, a dramatic rise in cerebellar cyclic GMP, weak inhibition of $(Na^+ + K^+)$-ATPase,

inhibition of neuronal 5HT uptake and induction of tremor. Sedation and/or convulsions have been described in animals following harmala alkaloids and other β-carboline derivatives.

Many tropical plants contain β-carboline structures with miscellaneous substituents and a varying degree of saturation. Carboxylic acid derivatives of β-carbolines are uncommon; however, the isolation of 1-methyl-β-carboline 3-carboxylic acid as the methyl ester from *Aspidosperma polyneuron* has been reported.

β-Carbolines of animal origin were subject to renewed interest in the early seventies when it was observed that tryptamine can easily react with N^5-methyl-tetrahydrofolic acid to form tetrahydro-β-carbolines. It was proposed that these tetrahydro-β-carbolines were involved in schizophrenia. Enzymes were involved in the reaction, but it appeared later that the fundamental coupling between tryptamine and enzymatically oxidized N^5-methyl-tetrahydrofolic acid was essentially a non-enzymatic Pictet-Spengler condensation.

The presence of tetrahydro-β-carbolines and β-carbolines *in vivo* in animals is pre-

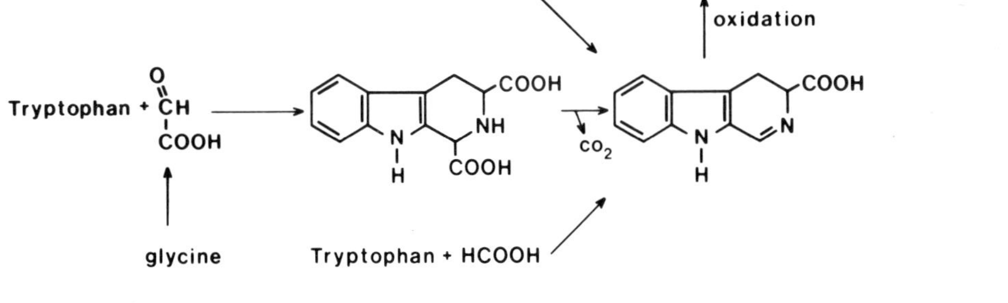

Fig. 3. Hypothetical pathways leading to β-carboline 3-carboxylic acid (β-CC). β-CC has low affinity for [³H]diazepam binding sites (Ki ≈ 15,000 nM).

carious on account of the ease with which tryptophan and tryptamine condense with aldehydes *in vitro* at physiological pH and temperature. Recent careful studies, however, have shown good evidence for the presence of low concentrations (20 ng g⁻¹ range) of norharman, harman, tetrahydroharman, and 6-hydroxytetrahydronorharman in normal blood platelets and/or brain tissue. Furthermore, it is particularly interesting that a 3-carboxylic acid derivative of tetrahydro-β-carboline has been isolated from cataractous human eye lenses, but not from normal lenses, as would be expected for an isolational artefact.

The biosynthetic pathways leading to β-carbolines are uncertain. There is general agreement, however, that tryptophan and tryptamine are involved. We have exemplified some hypothetical tentative pathways in Fig. 3.

Agonist/antagonist?

The possible presence of endogenous ligands for benzodiazepine receptors immediately poses another question; are benzodiazepines agonists or antagonists? Affinity binding studies cannot directly solve this problem since both agonists and antagonists usually compete for the same recognition site. It is known for several receptors and in particular for acetylcholine receptors, that agonists exhibit mixed type competitive inhibition of high affinity antagonist-sites, while antagonists exhibit pure competitive inhibition. Extrapolation of this observation would suggest that benzodiazepines are antagonists to 'agonists' of β-CCE-type, because β-CCE is a mixed type competitive inhibitor of [³H]flunitrazepam and [³H]diazepam binding in all rat brain areas except the cerebellum[3]. In addition, chloride increases [³H]diazepam binding; by analogy to the ion effects of high affinity receptor binding of other agonists and antagonists, respectively, this suggests that

benzodiazepines are antagonists[5]. Furthermore, when β-CCE was administered to mice in doses sufficient to block benzodiazepine receptors *in vivo* this compound did not antagonise pentazol induced convulsions, but reduced the protection against convulsions achieved by a benzodiazepine.

On the other hand, it is the rule rather than the exception that receptor antagonists enhance the binding capacity of the receptor. Prolonged benzodiazepine treatment did not, however, increase benzodiazepine binding; rather the opposite.

Conclusions

Benzodiazepine receptors are present in neuronal membranes as complexes, involving GABA receptors, chloride channels and/or other structures.

Benzodiazepines probably exert their effects by interacting with these receptors. There is no conclusive evidence for the presence of endogenous ligands, but it is likely that β-carboline 3-carboxylic acid may be part of either a small neurotransmitter-like or an extended neuromodulator-like ligand for these receptors. A biosynthetic pathway for β-carbolines should be demonstrated.

Reading list

1 Braestrup, C. and Squires, R. F. (1978) *Br. J. Psychiatr.* 133, 249–260
2 Braestrup, C., Nielsen, M., Krogsgaard-Larsen, P. and Falch, E. (1979) *Nature (London)* 280, 331–333
3 Braestrup, C., Nielsen, M. and Olsen, C. E. (1980) *Proc. Natl. Acad. Sci. U.S.A.* 77, 2288–2292
4 Costa, E. (1979) *Trends Pharmacol. Sci.* 1, 41–44
5 Costa, E., Rodbard, D. and Pert C. B. (1979) *Nature (London)* 277, 315–317
6 Haefely, W., Polc, P., Schaffner, R., Keller, H. H., Pieri, L. and Moehler, H. (1979) in *GABA-neurotransmitters* (Krogsgaard-Larsen, P., Scheel-Krueger, J. and Kofod, ' H. eds), pp. 357–375, Munksgaard, Copenhagen and New York
7 Sieghart, W. and Karobath, M. (1980) *Nature (London)* 286, 285–287
8 Young, W. S. III and Kuhar, M. J. (1979) *Nature (London)* 280, 393–395

9 Marangos, P. J., Paul, S. M. and Goodwin, F. K. (1979) *Life Sci.* 25, 1093–1102

10 Moehler, H., Polc, P., Cumin, R., Pieri, L. and Kettler, R. (1979) *Nature (London)* 278, 563–565

11 Squires, R. F. and Braestrup, C. (1977) *Nature (London)* 266, 732–734

12 Squires, R. F., Benson, D. I., Braestrup, C., Coupet, J., Klepner, C. A., Myers, V. and Beer, B. (1979) *Pharmacol. Biochem. Behav.* 10, 825–830

13 Tallman, J. F., Thomas, J. W. and Gallager, D. W. (1978) *Nature (London)* 274, 383–385

The use of interaction kinetics to distinguish potential antagonists from agonists

J. P. Raynaud, M. M. Bouton and T. Ojasoo

Roussel-Uclaf, 35 bld. des Invalides, Paris 75007, France.

The kinetics of the interaction between a ligand and the cytosolic steroid hormone receptor is an important factor in determining the nature and amplitude of the induced biochemical and biological responses and can be used as the basis of a screening system to distinguish potential antagonists from potent agonists for every class of steroid hormone.

The design of hormones of predetermined activity presupposes an understanding of their mode of action in order to be able to control *in vitro* the appropriate parameters involved in the *in vivo* response. It is generally accepted that, in the case of steroid hormones, the key event, in the absence of which no response occurs, is the interaction of the steroid hormone with a protein termed 'receptor' recovered in the high-speed supernatant (cytosol) of target organ homogenates. The complex formed between the steroid and receptor enters the nucleus where it binds to chromatin and triggers the subsequent course of events (RNA polymerase activity, RNA synthesis, protein synthesis...).

The interaction between steroid and receptor has been the object of innumerable studies, many of which have demonstrated that, if account is taken of modulating factors such as binding to specific plasma proteins, metabolism etc., the relative strength of the binding of a series of steroids to a receptor can be correlated with the end-response of the target organ[1]. Thus, for instance, the receptor binding of a radiolabelled potent progestin, promegestone, in rabbit or human uterus can be displaced by various competitors in an order related to their ability to induce endometrial proliferation in this organ[1]. However, although gross correlations can be drawn between binding and activity, it is difficult to draw a close relationship for several reasons; in particular, the relative binding affinity of a hormone measured under a fixed set of incubation conditions does not take into account the kinetics of the interaction between hormone and receptor and these kinetics, as shown below, can modulate the final response.

In our studies, we have postulated that if the complex between competing hormone (H) and cytoplasmic receptor (R_c) is more long-lived than that formed by the endogenous hormone, a potent agonist response will ensue. If, on the contrary, it is less long-lived, owing maybe to a con-

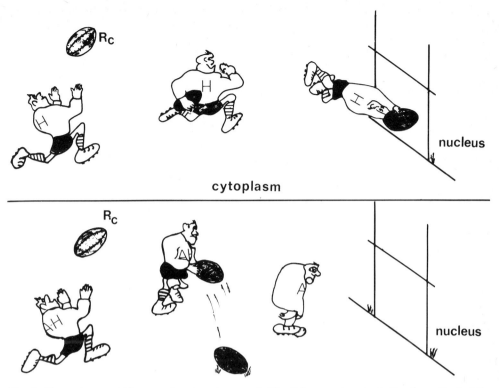

Fig. 1. The rugby man H(hormone) and rugby man A–H(anti-hormone) compete for the rugby ball R_c (receptor) in the field (cytoplasm). If the rugby man H is fast enough and catches the ball (high association rate), keeping it well tucked under his arm (slow dissociation rate, i.e. stable complex), he will score a try (translocation of complex into the nucleus). On the other hand, rugby man A–H may be fast enough to catch the ball (high association rate), but since he lets the ball easily slip out of his hands (high dissociation rate, i.e. unstable complex) few tries will be scored (no complex translocated). Although rugby man A–H cannot score many tries, he can effectively prevent rugby man H (e.g. the natural hormone) from doing so (anti-hormonal effect). Both the number of balls in the game and playing time are restricted.

formational misfit between H and R_c, only a partial response will be obtained and the action of the endogenous hormone will even be prevented if H can very readily occupy the available binding sites. In other words, antagonist activity will occur if the competing hormone has an association rate comparable to that of the endogenous hormone and if the dissociation rate of the complex thus formed is decidedly faster than the dissociation rate of the endogenous hormone complex (see Fig. 1).

This working hypothesis has been verified for steroids belonging to different hormone classes. For instance, estradiol derivatives with not very different association rates for R_c, but forming complexes with widely differing dissociation rates[2] (see Table I), can be classified into compounds able to maintain long-term estrogenic responses (e.g. ethynylestradiol and moxestrol) and compounds with limited intrinsic estrogenic activity (e.g. estriol and RU 16117) which can antagonize the action of estradiol. The slight modification in structure (replacement of an 11β-methoxy group by an 11α-methoxy group) converts a slowly-dissociating (i.e. potent) estrogen into a fast-dissociating (i.e. weak) estrogen with potential anti-hormonal activity. We have termed the competitor a 'potential' anti-hormone since the final activity of this compound

TABLE I. Interaction kinetics and biological potency of estradiol derivatives

	Association rate (k_{+1}) 0°C	Dissociation rate (k_{-1}) 25°C	Relative binding affinity		Uterotrophic activity	Anti-uterotrophic activity
			2 h, 0°C	5 h, 25°C		
Estradiol	20	30	100	100	1	—
Ethynylestradiol	46	15	112±5	245±13	3	—
Estriol	~7	115	15±2	4.3±0.5	0.005	10 v. 1
Moxestrol	~7	~4	12±1	122±11	10	—
RU 16117	~7	115	13±1	4.0±0.5	0.005	10 v. 1

Association rates (10^4 M^{-1} s^{-1}), dissociation rates (10^{-5} s^{-1}), relative binding affinities, relative uterotrophic potency (dose which increases uterine weight five-fold) and anti-uterotrophic activity (dose-ratio required to inhibit by 50% the increase in uterine weight induced by 0.3 μg estradiol) have been taken from Refs 2 and 3. Moxestrol: 11β-methoxyethynylestradiol; RU 16117: 11α-methoxyethynylestradiol.

will greatly depend upon the conditions under which it competes with the endogenous hormone[3]. Moderately large, well-spaced, doses will effectively prevent cytosolic estradiol binding and momentarily occupy most receptor sites forming fast-dissociating receptor complexes. Little complex is transferred into the nucleus, thus diminishing the natural replenishment phenomenon (estradiol induces the synthesis of its own receptor). The estrogenic response due to the anti-hormone will be weak and unsustained and that expected from the endogenous hormone will be prevented. However, if the cytosolic concentration of anti-hormone is permanently maintained at a certain level by frequent administration of low doses or by the use of implants etc., in spite of the rapid dissociation of the complex, there will always be present an adequate amount of anti-hormone to re-load the receptor and translocate it into the nucleus. An agonist response will thus ensue. In the case of RU 16117, it has been demonstrated that, with an appropriate administration schedule, it is possible to antagonize RNA polymerase activity (unpublished data), progesterone receptor synthesis[3] (estradiol induces the synthesis of this receptor), estradiol-induced uterine weight increase[3] and DMBA-induced tumours[4]. Furthermore, it has been recently shown that not only does RU 16117 dissociate faster than estradiol from the cytosolic receptor, but also from the nuclear receptor[5]. Thus not only is less complex translocated into the nucleus, but the nuclear retention of the complex is decreased. According to several authors[6], adequate nuclear retention is essential for long-term estrogenic responses.

In order to determine whether interaction kinetics similarly affect the responses observed with other hormone classes it was necessary to set up a system for establishing the kinetics of the $H.R_c$ inter-action compared with endogenous hormone complex formation without radio-labelling H. This was done by measuring the relative binding affinity (RBA) of H (compared with the endogenous hormone) under different incubation conditions[7]. These conditions were chosen in relation to the kinetics of interaction of the endogenous hormone with the receptor, the first incubation time reflecting primarily differences in association rates, the second incubation time, differences in dissociation rates. In general, if the RBA increases with incubation time, the complex formed dissociates slower than the endogenous hormone complex, i.e. H is a good agonist; if the RBA decreases with time, the complex is faster dissociating, i.e. H is a poor agonist and a potential antagonist. Indeed, as shown in Table I, the RBAs of moxestrol and RU 16117, which are similar after only 2 h incubation at 0°C, differ widely after 5 h incubation at 25°C. The RBA of moxestrol, like that of ethynylestradiol, has increased, whereas the RBA of RU 16117 has decreased comparably to that of estriol.

This principle was first validated in the case of androgens[8]. The substitution of a potent androgen (metribolone, 17β-hydroxy-17α-methylestra-4,9,11-trien-3-one) by a *gem*-dimethyl group at C-2 gave a compound, RU 2956, with anti-androgenic activity on androgen-induced prostate weight increase and mitotic activity. Whereas the RBA of metribolone for the cytosolic androgen receptor in rat prostate increased from 158 at 30 min to 203 at 2 h (ratio 1.3) the RBA of RU 2956 decreased from 55 to 14 (ratio 0.25). Non-steroid anti-androgens such as flutamide and RU 23908 (5,5-dimethyl-3-[4-nitro-3-(trifluoromethyl)-phenyl] imidazolidine-2,4-dione) were also found to compete for androgen receptor binding forming very fleeting complexes (decreasing RBAs) which might be the origin of their antagonist activity[8]. Other

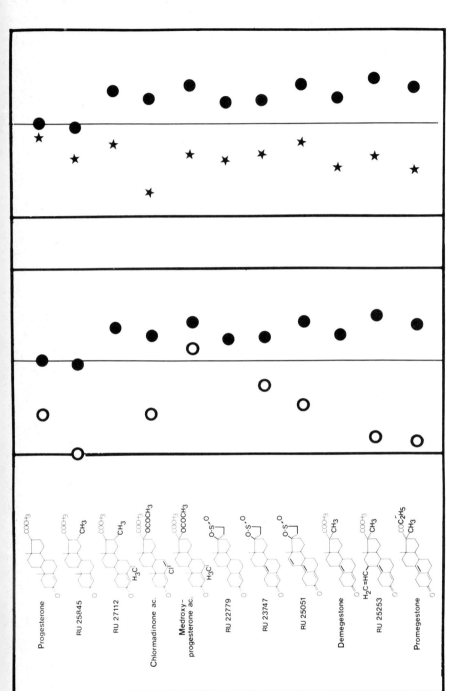

Fig. 2. Binding of progestins to the progestin receptor in rabbit uterus (●), the androgen receptor in rabbit uterus (○) and the glucocorticoid receptor in rat thymus (★). The ratios of the following RBAs are represented on a logarithmic scale: RBA of progestin binding measured after 24 h incubation over that measured after 2 h incubation (this ratio = 1 for progesterone); RBA of androgen binding measured after 2 h incubation over that measured after 30 min incubation (this ratio = 1 for testosterone); RBA of glucocorticoid binding measured after 24 h incubation over that measured after 4 h incubation (this ratio = 1 for corticosterone). A ratio <1 implies faster dissociation than the natural hormone, whereas a ratio >1 implies slower dissociation. Absolute RBA values for most compounds can be found in Refs. 11 and 12. The represented progestins are either anti-androgens or androgens, but all are apparently anti-glucocorticoids

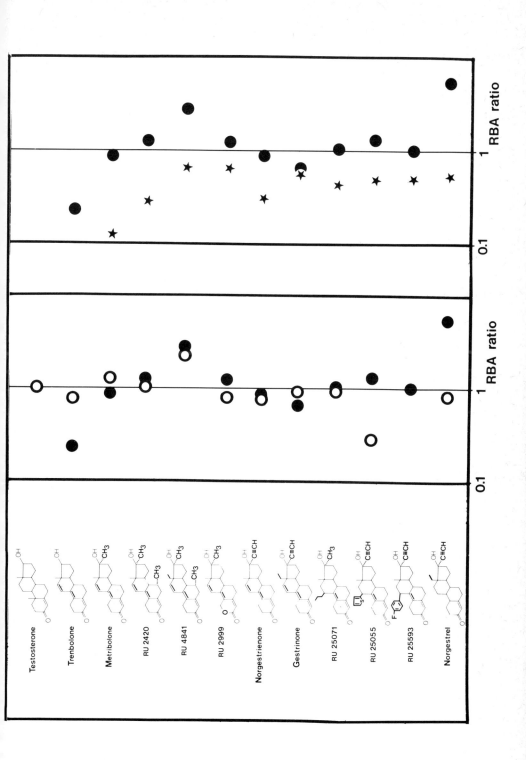

authors[9] have confirmed the validity of this principle for the accessory estrogenic binding of certain androgens, primarily androstanediols. These compounds compete appreciably for androgen receptor binding, but fleetingly for estrogen receptor binding. The high dissociation rates of the complexes they form with the estrogen receptor might explain their anti-estrogenic activity.

In the case of progestins, although a wide choice of compounds forming slowly-dissociating complexes is available (e.g. chlormadinone acetate, medroxy-progesterone acetate, promegestone etc. in Fig. 2), little success has been encountered in the synthesis of compounds forming fast-dissociating complexes. The RBA of gestrinone (13-ethyl-17β-hydroxy-18,19-dinorpregna-4,9,11-trien-20-yn-3-one) for the progestin receptor in rabbit uterus cytosol decreases from 76 to 48 on increasing incubation time at 0°C from 2 to 24 h. Although gestrinone has some anti-progestational activity in several tests[10], its intrinsic progestomimetic activity is too high to enable effective antagonism of progesterone action. Furthermore, this compound binds fairly markedly to the receptors of other hormone classes and thus has other biological actions.

Many progestins are potent anti-estrogens, but their anti-estrogenic activity, unlike that of androgenic androstanediols, does not involve competition for binding to the estrogen receptor, nor does it only involve an antagonist action on the replenishment of this receptor. Recent experiments have shown that it is correlated with the strength of the progestin binding[5]. On the other hand, the androgenic and anti-androgenic activities of progestins can be related to the kinetics of their interaction with the androgen receptor[11]. As shown in Fig. 2, progestins with a 17β-acetyl or a spiro-oxathiolane function at C-17, except for medroxyprogesterone acetate, form relatively fast-dissocia-

ting complexes with the androgen receptor. The ratio of the RBA recorded after long, over short, incubation times is mostly far smaller than 1. Medroxyprogesterone acetate has known androgenic properties, whereas compounds such as chlormadinone acetate, demegestone and promegestone can antagonize androgen action[11]. Progestins with a 17β-hydroxy function and either a 17α-methyl or ethynyl group nearly all form relatively slowly-dissociating complexes with the androgen receptor (except for the 11β-substituted steroids RU 25055 and RU 25593) and have androgenic activity (e.g. metribolone, RU 2999, norgestrel)[11].

All the tested progestins, whether progesterone or nortestosterone derivatives, formed fast-dissociating complexes with the cytosolic glucocorticoid receptor in rat thymus (Fig. 2) to the extent that it could be tentatively suggested that anti-glucocorticoid action might be a direct corollary of progestational activity. Indeed, several of these compounds can antagonize tyrosine-amino-transferase induction in rat hepatoma cells in culture[12], ACTH release in rat pituitary cells in culture (F. Labrie et al., to be published), ornithine decarboxylase induction (D. Lando, to be published) and the inhibitory effect of agonists on uridine incorporation into the ribonucleic acid of rat thymocytes[13].

The decrease in glucocorticoid binding concomitant with an increase in progestin binding would seem to imply a close relationship in the binding of a ligand to these two receptors. This is supported by the observation that non-progestational androgens do not seem to share the ability to bind to the same extent to the glucocorticoid receptor (see testosterone in Fig. 2, and unpublished data). On the other hand, some sistership would seem to exist between the androgen and mineralocorticoid receptors. It is now well established that several anti-aldosterone compounds compete not only for binding to the mineralocorticoid receptor (with decreasing RBAs with increasing incubation time), but also for testosterone binding to the androgen receptor (also with decreasing RBAs with increasing incubation time) and thereby possess anti-mineralocorticoid and anti-androgenic activities[11].

In conclusion, a simple screening system where the RBAs of a steroid for binding to several hormone receptors are measured under two sets of incubation conditions will enable the selection of compounds with potent agonist activity with regard to one or more receptors and furthermore will enable us to establish whether these same compounds (or others) may not also display anti-hormonal activity by virtue of their fleeting interaction with other receptors. In this way, the binding profiles of large numbers of steroids can be drawn and the data obtained used to establish closer structure–activity relationships for the design of new ligands and for the understanding of receptor binding requirements. Needless to say, however, the system will not select compounds which act as anti-hormones via another mechanism of action such as inhibition of hormone secretion, regulation of binding site concentration or receptor desensitization.

Reading list

1 Raynaud, J. P., Ojasoo, T., Bouton, M. M. and Philibert, D. (1979) in *Drug Design,* (Ariëns, E. J. ed.) Vol. VIII, pp. 169–214, Academic Press, New York.)

2 Bouton, M. M. and Raynaud, J. P. (1978) *J. Steroid Biochem.* 9, 9–15

3 Bouton, M. M. and Raynaud, J. P. (1979) *Endocrinology* 105, 509–515

4 Kelly, P. A., Asselin, J., Caron, M. G., Raynaud, J. P. and Labrie, F. (1977) *Cancer Res.* 37, 76–81

5 Raynaud, J. P. and Bouton, M. M. (1980) in Cytotoxic Estrogens in Hormone Receptive Tumors (Raus, J., Martens H. and Leclerq G.eds), pp 49–70, Academic Press, New York.

6 Clark, J. H., Paszko, Z. and Peck, E. J. Jr (1977) *Endocrinology* 100, 91–96

7 Raynaud, J. P. (1979) in (Jacob, J. ed.), *Advances in Pharmacology and Therapeutics* Vol. 1, pp. 259–278, Pergamon Press, Oxford

8 Raynaud, J. P., Bonne, C., Bouton, M. M., Lagacé, L. and Labrie, F. (1979) *J. Steroid Biochem.* 11, 93–99

9 Rochefort, H., Capony, F. and Garcia, M. (1979) *J. Steroid Biochem.* 11, 1635–1638

10 Sakiz, E., Azadian-Boulanger, G. and Raynaud, J. P. (1974) in *Proceedings, VI International Congress of Endocrinology, ICS 273,* pp. 988–994, Excerpta Medica, Amsterdam

11 Raynaud, J. P., Fortin, M. and Tournemine, C., (1980) *Actualités de Chimie Thérapeutique* 7, 293–318

12 Raynaud, J. P., Bouton, M. M., Moguilewsky, M., Ojasoo, T., Philibert, D., Beck, G., Labrie, F. and Mornon, J. P. (1980) *J. Steroid Biochem.* 12, 143–158

13 Dausse, J. P., Duval, D., Meyer, P., Gaignault, J. C., Marchandeau, C. and Raynaud, J. P. (1977) *Mol. Pharmacol.* 13, 948–955

Ubiquitous effects of the vitamin D endocrine system

Marian R. Walters, Willi Hunziker, and Anthony W. Norman

Department of Biochemistry, University of California, Riverside, California 92521, U.S.A.

Regulation of Ca^{2+} and PO_4^{2-} homeostasis by the vitamin D endocrine system involves complex interactions between both dihydroxylated metabolites: $1,25(OH)_2D_3$ and $24,25(OH)_2D_3$. New developments in this field include description of $1,25(OH)_2D_3$ receptors in many tissues not usually considered active in Ca^{2+} and PO_4^{2-} regulation. Additionally, $1,25(OH)_2D_3$ receptors behave differently than classical steroid hormone receptors.

The active forms of vitamin D, interacting with parathyroid hormone (PTH) and calcitonin, carefully regulate Ca^{2+} and PO_4^{2-} homeostasis throughout the body, with acute effects to maintain blood Ca^{2+} levels within certain defined limits[1-3]. A carefully regulated endocrine system similar to those of the other steroid hormones mediates the effects of the vitamin.

Synopsis of the vitamin D endocrine system

Key aspects of the vitamin D endocrine system are summarized in Fig. 1. The vitamin is a secosteroid in which the cyclopentanoperhydrophenanthrene B-ring is cleaved. It is obtained from dietary sources or from u.v.-directed conversion of 7-dehydrocholesterol in the skin. Because of the broken B-ring, in solution the A-ring of vitamin D and its metabolites undergoes rapid interconversions between two conformational states. Vitamin D (and its metabolites) is complexed in the blood to a specific α-globulin known as vitamin D binding protein for transport throughout the body. In the liver, the vitamin undergoes hydroxylation to 25-hydroxyvitamin D [$25(OH)D_3$], the predominant circulat-

ing form. Then in the kidney $25(OH)D_3$ undergoes an additional hydroxylation, resulting in conversion to the active hormone 1,25-dihydroxyvitamin D_3 [$1,25(OH)_2D_3$]. These hydroxylations occur via mitochondrial cytochrome P-450 containing hydroxylases. In addition, the kidney contains a 24-hydroxylase responsible for the formation of 24,25-dihydroxyvitamin D_3 [$24,25(OH)_2D_3$]. The specific functions of $24,25(OH)_2D_3$ are unknown, but there is evidence that $1,25(OH)_2D_3$ alone cannot account for all the functions of vitamin D and there is some direct evidence for $24,25(OH)_2D_3$ activity (see below). Studies are currently underway in this laboratory to assess the possible endocrine roles of this metabolite.

Due to the role of the kidney as the principal endocrine gland in the vitamin D system, it is also the primary site for control of the synthesis (availability) of the hormone(s). There is autoregulation in that $1,25(OH)_2D_3$ decreases 1-hydroxylase activity and increases that of the 24-hydroxylase. Additionally, PTH stimulates the 1-hydroxylase, but $1,25(OH)_2D_3$ inhibits this effect via feedback to the parathyroid gland. Other hormones regulate the activity of the 1-hydroxylase[1], however, some (e.g. estrogens) may act indirectly through changes in plasma Ca^{2+}.

After synthesis in the kidney in response to the Ca^{2+} 'demands' of the organism, $1,25(OH)_2D_3$ circulates to its target tissues to regulate Ca^{2+} and PO_4^{2-} absorption (intestine, kidney) or storage (bone mobilization/accretion). Under normal conditions, these effects in combination with those of PTH and calcitonin maintain plasma Ca^{2+} levels within a very narrow range (2.3–2.5 mM or 9–10 mg 100 ml^{-1} in the human).

Multiple target organs for $1,25(OH)_2D_3$

The primary target organs for $1,25(OH)_2D_3$: intestine, kidney, and bone, are involved in major aspects of Ca^{2+} and PO_4^{2-} regulation. Additionally, numerous other target tissues[4–7] have been recently defined on the basis of $1,25(OH)_2D_3$ receptor and/or vitamin D-dependent CaBP (Ca^{2+} binding protein) localization. In some cases $1,25(OH)_2D_3$ functions in these

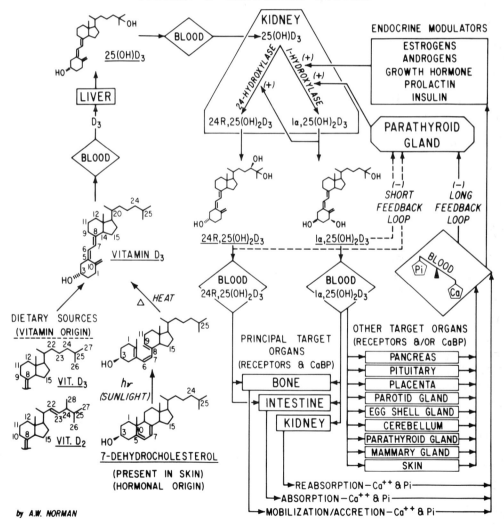

Fig. 1. Synopsis of the vitamin D endocrine system.

tissues can be readily postulated. Thus $1,25(OH)_2D_3$ receptors in the parathyroid gland may function in the feedback loop for control of PTH secretion and/or synthesis to ultimately regulate the 1-hydroxylase activity. In placenta, mammary gland, and egg shell gland, $1,25(OH)_2D_3$ receptors may produce biological responses which enhance Ca^{2+} transport to the fetus, milk, and egg shell, respectively. In other tissues, considerations of function are only speculative at this time. Thus $1,25(OH)_2D_3$ receptors in the pituitary may relate to unknown $1,25(OH)_2D_3$ releasing factors/ inhibitors. In the pancreas, there may be some relation to Ca^{2+} regulation of insulin secretion. In skin, $1,25(OH)_2D_3$ receptors may regulate some aspect of the B-ring cleavage, e.g. u.v. availability via melanization, product storage, or substrate availability. Importantly, the widespread tissue distribution of $1,25(OH)_2D_3$ receptors is not dissimilar to that of other specialized steroid hormones such as estrogen and progesterone, and definition of additional target tissues will probably continue.

Receptors for $1,25(OH)_2D_3$: recent developments

Following the initial recognition that vitamin D acts via an endocrine mechanism similar to that of other steroid hormones, cytosol $1,25(OH)_2D_3$ receptors were found in many tissues, as predicted a priori by analogy to the steroid hormones. However, the detection of these cytosol $1,25(OH)_2D_3$ receptors required buffers with intermediate or high ionic strengths, which can extract even occupied receptors from the nuclei.

Re-evaluation of the apparent lack of cytosol $1,25(OH)_2D_3$ receptors in the low salt buffer routinely used with other steroid hormones produced unexpected, but potentially very significant, results[8]. Under these conditions, 90% of the unoccupied $1,25(OH)_2D_3$ receptors were localized in crude chromatin preparations. The co-identity of these receptors with 'cytosol' receptors obtained after homogenization in high salt buffers was confirmed by their biochemical characteristics. This in vitro nuclear association appeared to be a general phenomenon, since the majority of the unoccupied $1,25(OH)_2D_3$ receptors were found in crude chromatin preparations in all tissues examined (Table I). Importantly, the proportion of unoccupied $1,25(OH)_2D_3$ receptors associated with nuclear components varied inversely with the ionic strength of the homogenization buffer; but at infinitely small dilutions (i.e. in vivo) little or no receptor was present in the cytosol even in the high salt buffer. Additionally, the presence of unoccupied $1,25(OH)_2D_3$ receptors has been subsequently confirmed in purified nuclei. Taken collectively, these data suggest that unoccupied $1,25(OH)_2D_3$ receptors are largely, but not necessarily exclusively, localized within nuclei in vivo. Thus they may be more closely related to triiodothyronine receptors than previously recognized.

Another aspect of the $1,25(OH)_2D_3$ receptor system which has recently progressed is the study of occupied $1,25(OH)_2D_3$ receptors and their correlations to hormonal effects. By blocking ligand binding to unfilled receptors at 4°C by prior incubation with TPCK (L - 1 - tosylamide - 2 - phenyl - ethylchloromethyl ketone), we have developed a quantitative exchange assay for measuring in vivo occupied as well as unoccupied $1,25(OH)_2D_3$ receptors[9]. This exchange assay is already providing important insights into the physiological and biochemical roles of $1,25(OH)_2D_3$ and its receptors in inducing hormonal responses (Fig. 2).

$24,25(OH)_2D_3$: an essential component of the vitamin D endocrine system

The renal 24-hydroxylase is carefully regulated and circulating levels of

TABLE I. Unoccupied 1,25(OH)$_2$D$_3$ receptors are found associated with the crude chromatin fraction in many target tissues.

Tissues	1,25(OH)$_2$D$_3$ receptors in chromatin (%)
Intestinal mucosa (chick, rat)	90, 92
Parathyroid gland (chick)	82
Kidney (chick, rat)	61, 72
Pancreas (chick)	77
Testes (rat)	61
Bone cell culture (mouse)	84

24,25(OH)$_2$D$_3$ are very high. These observations have led to the speculation that this vitamin D metabolite may have important hormonal functions which are distinct from, but integrated with, those of 1,25(OH)$_2$D$_3$. To date this concept is disputed, but strong evidence is accumulating for a biological role for 24,25(OH)$_2$D$_3$[10]. In the chick parathyroid gland, only the simultaneous administration of 24,25(OH)$_2$D$_3$ plus 1,25(OH)$_2$D$_3$ reverses the hypertrophy and hyperplasia induced by vitamin D deficiency. In the dog, 1,25(OH)$_2$D$_3$ and 24,25(OH)$_2$D$_3$ have acute effects on the secretion of immunoreactive PTH; but 24,25(OH)$_2$D$_3$ alone reduces iPTH levels when given chronically to uremic, hyperparathyroid dogs[11]. In the rachitic chick only 24,25(OH)$_2$D$_3$ plus 1α(OH)D$_3$ (converted to 1,25(OH)$_2$D$_3$ in the liver) induces bone healing[12]. Thus functional roles have been implicated for 24,25(OH)$_2$D$_3$ in both the parathyroid gland and in bone. Because the Ca^{2+} and PO$_4^{2-}$ demands, and hence hormonal regulation, may vary greatly throughout the life cycle, chicks were raised from hatching to maturity on a vitamin D deficient diet supplemented with vitamin D, 1,25(OH)$_2$D$_3$, 24,25(OH)$_2$D$_3$ or combinations thereof[10]. The results of this experiment demonstrate that (a) normal egg hatchability requires both 24,25(OH)$_2$D$_3$ and 1,25(OH)$_2$D$_3$, (b) the integrity of bone structure requires both 24,25(OH)$_2$D$_3$ and 1,25(OH)$_2$D$_3$, and (c)

both 24,25(OH)$_2$D$_3$ and 1,25(OH)$_2$D$_3$ are required to maintain Ca^{2+} and PO$_4^{2-}$ homeostasis within the normal range in the chick[10]. Taken collectively, these results emphasize the importance of both 24,25(OH)$_2$D$_3$ and 1,25(OH)$_2$D$_3$ in regulating various aspects of Ca^{2+} and PO$_4^{2-}$ homeostasis in multiple tissues. Additionally, these conclusions have important biochemical and clinical ramifications. For example, in renal osteodystrophy where osteomalacia is not corrected by 1,25(OH)$_2$D$_3$ treatment alone, the possibility remains that 24,25(OH)$_2$D$_3$ administered with 1,25(OH)$_2$D$_3$ may induce improvement.

Conclusions

The vitamin D endocrine system plays a major role in maintaining plasma Ca^{2+} and PO$_4^{2-}$ homeostasis. In addition, these secosteroid hormones are important in other aspects of Ca^{2+} physiology, including providing Ca^{2+} for bone mineralization, milk production, and fetal growth. They may also be involved in more precise events, such as regulation of insulin secretion. 1,25(OH)$_2$D$_3$ is known to be involved in all these phenomena due to evidence of its

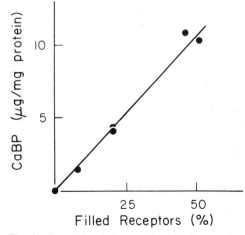

Fig. 2. Correlation between 1,25(OH)$_2$D$_3$ receptor occupancy (2 h post-injection) and subsequent biological response e.g. the appearance of a calcium binding protein (CaBP) (20 h post-injection) in the chick intestinal mucosa.

action after treatment *in vivo*, receptor localization, and the presence of vitamin D dependent CaBP. Additionally, although the specific role(s) of $24,25(OH)_2D_3$ remains to be elucidated, this metabolite has been shown to have hormonal functions which are distinct from, but integrated with, those of $1,25(OH)_2D_3$. Recent studies of $1,25(OH)_2D_3$ receptors have produced unexpected results with important biochemical and physiological ramifications.

Reading list

1 Norman, A. W. (1979) in *Vitamin D: The Calcium Homeostatic Steroid Hormone*, Academic Press, New York
2 Haussler, M. R. and McCain, T. (1977) *N. Engl. J. Med.* 297, 974–983 and 1041–1050
3 DeLuca, H. F. and Schnoes, H. K. (1976) *Ann. Rev. Biochem.* 45, 631–666
4 Abstracts of the 62nd Meeting of the Endocrine Society (1980) *Endocrinology* 106, 22A, 23A, 322A, 861A
5 Pike, J. W., Gooze, L. L. and Haussler, M. R. (1980) *Life Sci.* 26, 407–414
6 Stumpf, W. E., Sar, M., Reid, F. A., Tanaka, Y. and DeLuca, H. F. (1979) *Science* 206, 1188–1190
7 Christakos, S. and Norman, A. W. (1980) *Fed. Proc.* 39, 1554A
8 Walters, M. R., Hunziker, W. and Norman, A. W. (1980) *J. Biol. Chem.* 255, 6799–6805
9 Hunziker, W., Walters, M. R. and Norman, A. W. (1980) *J. Biol. Chem.* 255 (20), 9534
10 Norman, A. W., Henry, H. L. and Malluche, H. H. (1980) *Life Sci.* 27, 229–237
11 Canterbury, J. M., Gavellas, G., Bourgoignie, J. J. and Reiss, E. (1980) *J. Clin. Invest.* 65, 571–576
12 Ornoy, A., Goodwin, D., Noh, D. and Edelstein, S. (1978) *Nature (London)* 276, 517–519

The haemoglobin molecule: is it a useful model for a drug receptor?

Peter J. Goodford

Wellcome Research Laboratories, Beckenham, Kent, England.

One of the interesting properties of bio-logically-active compounds is their selec-tivity. A naturally-occurring hormone or neurotransmitter will influence certain functions in a living organism specifically, while other functions in the same organism are relatively unaffected. Even the simplest forms of life rely on this selectivity as the chemical basis for their feedback processes. In higher forms such as whole animals, very complex and subtle controls have been developed which exploit a variety of different mechanisms in order to optimize the discriminatory powers of the controlling molecules.

Turning away from natural effector substances to modern drug molecules, attempts have been made to use some of the same mechanisms in order to improve the selectivity of drugs. Natural hormones in the body for instance, are often delivered by physiological mechanisms close to their intended site of action, so that they have the best possible chance of acting at the right location and nowhere else. Thus, nerve terminals release minute quantities of neurotransmitter in the closest proximity to the cells which the nerves control. Similarly, although with much less discrimination, attempts are made to ensure that drugs reach the intended parts of the body. Again there are enzyme mechanisms which decompose neurotransmitters after they have done their job, so that the overall level of transmitter does not rise in the body after a burst of nervous activity. This increases selectivity because the blood concentra-tion of the biologically active substance never exceeds an acceptable threshold level, and analogous properties have been conferred on drug molecules. For example the chemical structure of the drug decamethonium was modified to give suxamethonium, which is degraded by the enzyme cholinesterase in the blood so that excessive blood concentrations are never reached.

Such processes controlling the delivery and fate of molecules in the body, may be called distribution mechanisms. If they could provide adequate selectivity by themselves, only one type of natural effector molecule and one receptor might be needed to control all biological pro-cesses. The very fact that different effectors and receptors have evolved, shows that distribution mechanisms alone cannot provide such precise discrimina-tion as the drug–receptor interaction. This review article therefore deals with a model receptor, haemoglobin, and the factors which determine the specificity of its reactions with model drugs.

Receptors

Biological macromolecules often consist of many thousands of atoms, and yet the chemical structure and biological properties of each macromolecule are precisely defined. This precision is achieved by the use of only a few fundamental building blocks which are the modules from which the whole structure is constructed. In the case of the proteins these modules are the 20 naturally-occurring amino acids, and they are joined end to end in a long chain to form the protein macromolecule. The sequence in which the different amino acids occur determines the chemical structure, and very largely determines the chemical and biological properties of the protein as well. This follows since any particular sequence of amino acids does not exist in solution as an extended chain but folds into a more compact structure, just as a long strand of wool may be wound up into a ball. The chemical sequence controls the folding, and the folding controls the biological properties because it brings certain crucial parts of the macromolecule into close juxtaposition. Perhaps surprisingly, this final folded protein structure is very well defined.

It is widely accepted that such proteins can discriminate most effectively in their interactions with small molecules. When the interaction is an essential link in a natural control mechanism, the macromolecule is commonly termed a 'receptor'. When the interaction is coincidental, although perhaps still very specific, the macromolecule is said to provide a 'binding site'. When the interaction is part of a metabolic or degradative or synthetic process, the macromolecule is an 'enzyme'. Moreover, it is implicit in the definition of a receptor that the macromolecule not only interacts selectively with the incoming 'effector' molecule, but also transmits information to the next stage of the control system.

In essence, then, it would seem that a receptor may be defined on the basis of two properties. First, the ability to react rather specifically with a small effector molecule. Second, an ability to respond to the interaction in a way which will influence the future behaviour of the biosystem. Such a response might be observed experimentally as a change in the local electrical potential, or as a change in the local concentration of a 'second messenger' molecule. Thus, when glucagon interacts with liver cells one of the first effects to be observed is a change in the local concentration of cyclic AMP, owing to the activation of adenylate cyclase, although the detailed molecular events which occur between the glucagon interaction at its receptor and the observed increase of cyclic AMP concentration, are almost completely unknown.

Many pharmacologists would also restrict the term 'receptor' to membrane-bound macromolecules, in contrast to macromolecules in solution. This is an important distinction because no detailed information is yet available on the structure of membrane-bound receptors, and it is tempting to speculate that they may have special properties which do not occur in the soluble proteins whose structure has been determined so far. Be this as it may, we may safely conclude that those general properties and functions which have been observed in soluble proteins, may also be displayed with appropriate modifications by membrane-bound receptors. Soluble proteins should therefore provide interesting information of relevance to receptors in general.

The structure and function of haemoglobin

Haemoglobin is a soluble protein which

occurs in the red blood cells, and reacts reversibly with oxygen[2]. It takes up oxygen as blood passes through the lungs, where the oxygen partial pressure is high, and releases the oxygen as the blood passes through the tissues of the body where the partial pressure is lower. However, this release is not a straightforward process, and under normal physiological conditions it is necessary for a small effector molecule, 2,3,-diphospho-glycerate (DPG), to react with the haemoglobin as it gives up its oxygen[5,6,9]. The overall process may be represented, in a simplified way, by the Eqn (1).

$$Hb(O_2)_4 + DPG = Hb.DPG + 4O_2 \quad (1)$$

This is, to a greater or lesser extent, analogous to a receptor system in which DPG is the effector molecule and oxygen is the second messenger. Haemoglobin is the receptor and is, moreover, a substance whose properties are extremely well known. The protein sequence has been determined, and high resolution X-ray crystallographic methods have been used

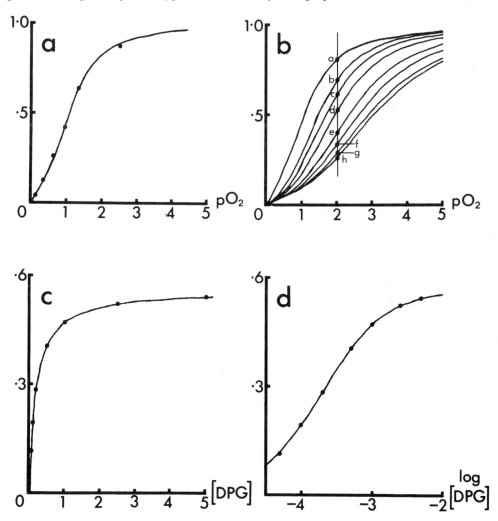

Fig. 1. This shows how a set of oxygen dissociation curves for haemoglobin can be reinterpreted as a pharmacological log concentration–response curve. See text.

to determine the three-dimensional structure of the molecule[10]. Furthermore, the uptake and release of the 'second messenger' oxygen can be easily measured, and so the system is readily available for experimental study.

A dissociation curve is the traditional way in which the physiological properties of haemoglobin are described, under any particular set of experimental conditions. This is a graph which relates the partial pressure of oxygen (pO_2) to the proportion of oxyhaemoglobin $Hb(O_2)_4$. In pharmacological nomenclature the dissociation curve relates the concentration of second messenger (oxygen) to the amount of second messenger in the receptor system, but this is not very helpful because the latter quantity cannot be readily determined for a pharmacological receptor. The pharmacologist would instead plot the response (i.e. the amount of oxygen liberated) against the concentration of effector (DPG).

Fig. 1 shows how these different representations of the experimental data can be related to each other. In Fig. 1a there is a single haemoglobin dissociation curve, fitted through six experimentally determined observations. A set of such dissociation curves is shown in Fig. 1b, although the experimental points are here omitted and the curves were calculated by theory in order to eliminate the additional complications of experimental error. The left-hand curve in Fig. 1b shows the dissociation curve in the complete absence of effector, and the seven successive curves to the right show what would happen in the presence of 0.05, 0.1, 0.2, 0.5, 1, 2.5 and 5 mM DPG. Now consider the points **a–h** where the different dissociation curves cut the vertical line which corresponds to a fixed partial pressure of oxygen. Each point corresponds to one concentration of the effector substance, DPG. At point **a** no DPG is present, and the haemoglobin is largely saturated with

oxygen. At point **b** there is 0.05 mM DPG, and a small amount of oxygen has been displaced. More oxygen has been displaced at point **c** and more DPG is present, and so on up to point **h** which corresponds to the highest DPG concentration studied and the greatest amount of oxygen liberated. These points in fact provide the basis for constructing the dose–response curve shown in Fig. 1c where the amount of oxygen liberated at constant partial pressure is plotted as the response against the corresponding concentration of DPG in mmol l^{-1}. Fig. 1d shows the corresponding plot of response against log concentration, and thus the physiologists' dissociation curve has been transformed into the traditional pharmacologists' log concentration–response curve. To this extent at least, haemoglobin provides a receptor model which is amenable to quantitative interpretation.

Structure–function relationships

The function of haemoglobin as an oxygen-carrier has been related to its observed macromolecular structure. It consists, in fact, of four separate protein chains which are folded up into four distinct globular subunits surrounding haem groups containing iron atoms. The protein chains do not all have the same sequence, and the different sequences are denoted by the Greek letters α, β, γ, and so on. Normal haemoglobin is a tetramer assembled from two α and two β chains packed together with the four iron atoms at the apices of a tetrahedron (Fig. 2). The four subunits nestle very closely together in the presence of oxygen, with an oxygen molecule combined at each iron atom. This is oxyhaemoglobin. In the complete absence of oxygen however, there is a structural change as the oxygen is released from the iron atoms, and the four subunits repack into a slightly more open assembly (Fig. 2). With this deoxy structure the affinity of the iron atoms for

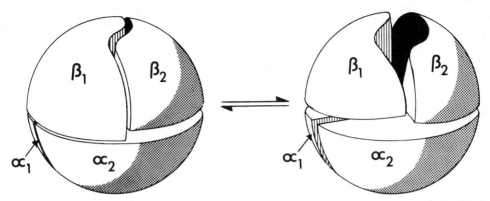

Fig. 2. The four subunits pack closely together in oxyhaemoglobin, but a crevice between the β-subunits opens in deoxyhaemoglobin.

oxygen is lowered which is why it is deoxyhaemoglobin, in contrast to the oxy form which has a higher oxygen affinity[2].

It should be noted in Fig. 2 that the crevices between the subunits are wider in deoxyhaemoglobin, and one crevice between the two β subunits opens particularly wide. Molecular models show that the DPG molecule fits this crevice almost perfectly, and X-ray crystallographic studies[1] have shown that this is the receptor site where DPG combines. The interaction is, moreover, selective for deoxyhaemoglobin because the site virtually disappears in the oxy form (Fig. 2), and this selectivity for deoxyhaemoglobin is essential if a compound is to influence the oxy–deoxy equilibrium and thereby promote the liberation of oxygen.

The chemical structure of the receptor cleft has been determined by X-ray crystallography in deoxyhaemoglobin with DPG present, and in oxyhaemoglobin without. Both are shown in Fig. 3. The receptor site is perfectly symmetrical because it lies between the two identical β subunits. It contains eight basic groups, four of which are $-NH_2$ and four of which are histidine residues. The natural effector DPG contains eight potentially acidic oxygen atoms, three on each phosphate and two on the carboxy group. When it combines with the receptor on deoxyhaemoglobin, seven of these eight acidic groups interact with seven of the eight basic residues (Fig. 3), and the overall fit of this natural effector to its receptor is impressive although the eighth potential acid–base interaction does not occur[1].

It may be asked how an interaction at the cleft between the β subunits can change the oxygen affinity of the iron atoms, since they are some distance away. In thermodynamic terms DPG shifts the equilibrium between the oxy and deoxy forms of the protein. In structural terms the X-ray crystallographic evidence shows that there are small differences throughout the haemoglobin molecule between the forms, and small structural changes adjacent to the iron atoms alter their oxygen affinities. Thus the whole process from receptor interaction to 'second messenger' liberation has been established for the haemoglobin system. The structure of the protein has been related to its function, and this is a necessary prerequisite before haemoglobin can be used as a model at the molecular level for the interpretation of receptor properties.

Drug design by the method of receptor fit

Since the structure and function of the DPG site are understood at the molecular level, it is natural to enquire if this information can be used for the design of novel compounds to interact at the site and

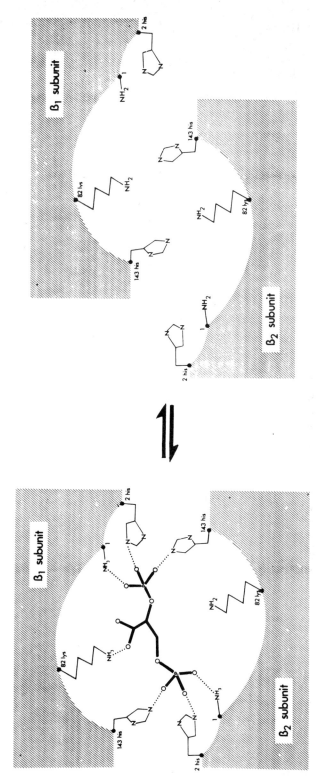

Fig. 3. The chemical structure of the receptor cleft in deoxyhaemoglobin with DPG, and in oxyhaemoglobin without.

thereby liberate oxygen[7]. In pharmacological terms such compounds could be described as 'agonists' interacting with a receptor and promoting the release of a second messenger, oxygen. Moreover, such an approach to drug design starting from the unknown structure of a receptor, should provide significant advantages over more conventional approaches. The investigator would not be constrained to begin his drug-design strategy from the chemical structure of a known natural effector. Instead of looking for compounds which are chemically related to DPG, he could look directly for compounds to react with the receptor site itself. Since this contains $-NH_2$ groups, which can react chemically with aldehydes, he might investigate compounds containing aldehyde groups at the correct spatial separation to interact optimally with the $-NH_2$ groups of the receptor in its deoxy form. The chosen compounds should not of course interact with oxyhaemoglobin, because a specific interaction with deoxyhaemoglobin is required.

Examination of Fig. 3 shows that the $-NH_2$ groups on residue lysine 82 are separated by approximately the same distance in both oxy- and deoxyhaemoglobin, so that any compound designed to interact with them might not differentiate very effectively between the two forms of the protein. However, the other two $-NH_2$ groups at the receptor move far apart in oxyhaemoglobin (Fig. 3) but are relatively close together in deoxyhaemoglobin. Dialdehyde compounds were therefore designed to fit the latter structure[3], and it was hoped to achieve selectivity for the deoxy form by choosing molecules which were not long enough to bridge the gap between the chosen $-NH_2$ groups in their widely separated positions on the oxy structure.

A series of bibenzyl dialdehydes related

to compound I had the appropriate properties.

I

Benzene rings were deliberately incorporated into this chemical structure, because benzenoid substitution should allow related compounds to be made in order to optimize the physicochemical properties of the molecule. Benzenoid substitution could, for example, be used to increase or decrease the chemical reactivity of the aromatic aldehyde groups. It could also influence lipophilicity, and since compound I was insoluble a solubilizing group was introduced to give compound II.

II

It was hoped that the carboxylic sidechain in this compound would not only increase solubility, but would also facilitate interaction with the receptor site by reacting with the protein like the carboxy group of DPG. The predicted mode of interaction is shown in Fig. 4a, and compound II promoted the liberation of oxygen although it was less active than DPG. Moreover it is known that aromatic aldehydes can react with bisulphite to form derivatives such as compound III.

III

When this compound was modelled into the DPG receptor site, it not only fitted

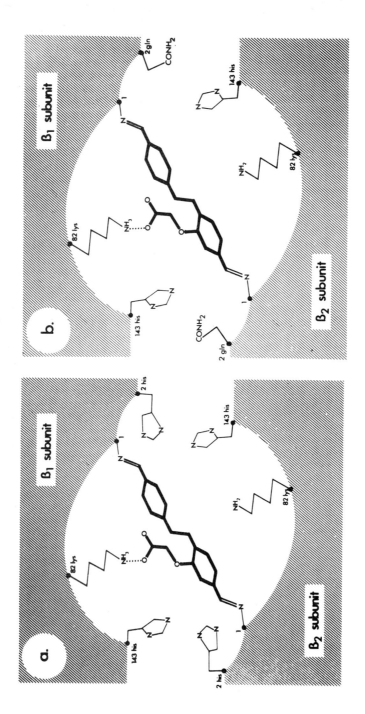

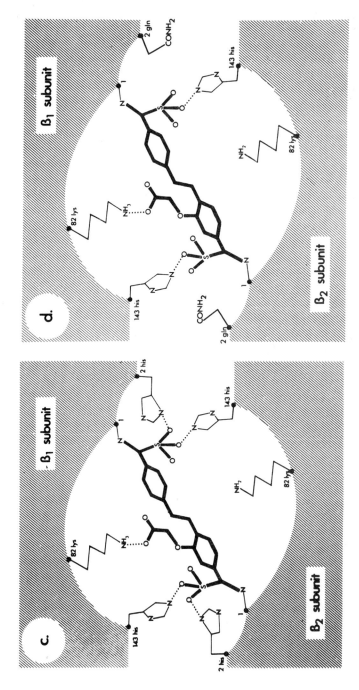

Fig. 4. The predicted interactions of the novel synthetic compounds II and III, with human and horse haemoglobin. See text.

very well but the sulphur–oxygen bonds pointed in just the right directions to interact with the histidine residues at the receptor. We therefore concluded that compound III should be more potent than compound II, because it could make more interactions with the receptor site[3]. The predicted mode of interaction is shown in Fig. 4c, and when compound III was tested it actually turned out to be more potent than DPG itself.

These simple findings with haemoglobin suggest that a knowledge of the structure and function of a receptor can be used in favourable cases to design novel types of biologically active agonist compounds. It seems likely that 'drug design by the method of receptor fit' may become widely used in the years ahead, as the structure of more receptors is established experimentally. Once again haemoglobin provides an interesting model.

Designing selective compounds

Modern drugs are selective. They have to kill microbes but not man, or affect smooth muscle but not the heart although both the tissues may have receptors of closely related structure. The question therefore arises whether a knowledge of the receptor structures can be used in order to design compounds which will affect one but not the other. Haemoglobin may be used as a model to investigate this type of requirement, because a wide range of different haemoglobins have been prepared from different biological sources. Every animal species has its own variety of haemoglobin with the same general tetrameric structure, but having a greater or lesser proportion of the normal amino acid residues replaced by different amino acids. Moreover many different types of human haemoglobin are known. Some occur in large population groups and others have only been found in single individuals, but overall they provide a wide test bed of different yet related receptor models, whose properties can be compared and contrasted.

For present purposes attention must be focused on those haemoglobin variants which differ from normal adult human haemoglobin in the region of the DPG receptor site. In horse haemoglobin for instance, two of the basic histidine residues at the receptor site are replaced by non-basic glutamine residues which would interact much more weakly with the acidic oxygen atoms of DPG. One might therefore predict that DPG should have a reduced potency against horse haemoglobin, and this is in fact the case. For compound II on the other hand, the prediction is (Fig. 4a) that the molecule only interacts with $-NH_2$ groups, and so the difference between horse and human haemoglobin at the DPG site should not matter. The predicted interaction between compound II and horse haemoglobin is shown in Fig. 4b, and is essentially the same as the predicted interaction with human haemoglobin in Fig. 4a. Moreover, the experimental evidence is again compatible with this prediction since the potency of compound II is not significantly different when tested on human and horse haemoglobins[4].

The prediction for compound III and human haemoglobin is shown in Fig. 4c, and may be compared with the horse haemoglobin prediction in Fig. 4d. In this case one would expect the potency on horse haemoglobin to be diminished, and the experimental results also confirm this prediction[4]. Moreover, similar observations have been made on two types of human foetal haemoglobin, on a modified haemoglobin which occurs in the blood of untreated diabetic patients, and on a human mutant haemoglobin. A total of six different receptors have been tested with three compounds (DPG, compound II and compound III) giving eighteen different compound–receptor interactions, and when the results were finally brought together it was found that the potency of each compound in releasing oxygen from any particular haemoglobin could be simply and quantitatively predicted[4]. Essentially, the free energy of each interaction was determined by the predicted number of salt bridges and covalent bonds between compound and receptor. Of course this may be a special case and such a simple relationship may not be generally applicable. Indeed the relationship could still be coincidental, although this is statistically unlikely. It is also possible that the compounds may interact at other parts of the haemoglobin molecule, as well as the DPG site itself. Nevertheless the present findings suggest that the method of drug design by receptor fit can be used in favourable cases in order to design compounds which interact selectively with one receptor in preference to another similar one.

Three-state models

So far we have discussed how haemoglobin resembles a drug-receptor system which can exist in two distinct 'states'. According to this interpretation the second messenger oxygen is liberated when the equilibrium between the states is shifted from oxy- to deoxy-haemoglobin, because deoxyhaemoglobin has a lowered oxygen affinity. DPG interacts selectively with the deoxy form, shifting the oxy-deoxy equilibrium and thereby promoting oxygen liberation. However, it turns out that this simple two-state interpretation is inadequate when the detailed observations on the haemoglobin/DPG/oxygen system are carefully analysed.

Fortunately a simple extension of the model can explain the detailed findings. In the haemoglobin system, the first state is oxyhaemoglobin. This is a form of the protein having high affinity for oxygen but very low affinity for DPG. The second state is deoxyhaemoglobin with a lower oxygen affinity and a DPG receptor site which gives it a high DPG affinity. Up till now it has been tacitly assumed that the presence of DPG at its receptor simply changes the mass action equilibrium between the two states, but has no direct influence on the properties of the deoxy-haemoglobin itself. This tacit oversimplification must now be rejected. Instead, it is necessary to assume that DPG fulfils two distinct roles[8]. It not only shifts the mass action equilibrium, but changes the properties of deoxyhaemoglobin when it reacts by further lowering the already low oxygen affinity of the deoxy protein. Hence one may say that there are three haemoglobin states:

> Oxyhaemoglobin with high oxygen affinity
> Deoxyhaemoglobin with lower oxygen affinity
> Deoxyhaemoglobin (combined with DPG) with still lower affinity.

Such a three-state model is intuitively satisfying and physically reasonable. The properties of molecules are normally altered when they interact with each other, and it was an oversimplification to assume that the oxygen affinity of deoxyhaemoglobin did not change at all after DPG interaction. Moreover the three-state model can be extended to DPG analogues such as compounds II and III, and to other compounds which occur naturally in red blood cells such as adenosine triphosphate and inositol penta-phosphate[8]. Cutting a long story short, it turns out that these compounds all differ from each other in two distinct properties, both of which determine their overall ability to influence the haemoglobin-oxygen system. First, each compound has its own particular affinity for deoxy-haemoglobin which determines its action on the oxy–deoxy equilibrium. Second, the extra lowering of the oxygen affinity of deoxyhaemoglobin varies from compound to compound. In other words, the oxygen affinity of the third state depends on the particular agonist which is used.

This type of distinction between agonists has not yet been detected in conventional pharmacological systems.

Conclusions

The relationships between protein structure and function are better understood for haemoglobin than for any comparable system. They provide the background knowledge which allows haemoglobin to be used as a model for pharmacological receptors. The model is, of course, very simple, and many pharmacological receptors may have additional properties which are not displayed by haemoglobin itself. Nevertheless, the haemoglobin structure has been used to design biologically active, novel compounds by the method of receptor fit, and to predict the selectivity of such compounds. Moreover, their potency can be quantitatively measured, and can be roughly interpreted on the basis of a simple two-state model.

The need for the three-state model is, perhaps, the most exciting finding in the present work. It suggests that compounds may not only differ in their affinities for a given receptor, but that they may also differ in the nature of the effects which they produce there. In the case of haemoglobin they differ in their ability to depress the already low oxygen affinity of deoxyhaemoglobin. In contrast to this finding, however, the available experimental evidence on conventional pharmacological receptors can be explained by only two states.

It is always possible of course, that haemoglobin is a very unusual protein in requiring three states to explain its mode of action. However, an alternative and more challenging interpretation is also available. Proteins in general, and pharmacological receptors in particular, may really exist in multiple states. Perhaps these have not yet been detected in conventional pharmacological systems, because the appropriate experiments have not been designed. Such experiments should now be planned and carried out.

Reading list

1 Arnone, A. (1972) *Nature (London)*, 237, 146–148
2 Baldwin, J. M. (1975) *Prog. Biophys, Molec. Biol.* 29, 225–320
3 Beddell, C. R., Goodford, P. J., Norrington, F. E., Wilkinson, S. and Wootton, R. (1976) *Br. J. Pharmacol.* 57, 201–209
4 Beddell, C. R., Goodford, P. J., Stammers, D. K. and Wootton, R. (1979) *Br. J. Pharmacol.*, 65, 535–543
5 Benesch, R. and Benesch, R. E. (1967) *Biochem. Biophys. Res. Commun.* 26, 162–167
6. Douglas, C. G., Haldane, J. S. and Haldane, J. B. S. (1912) *J. Physiol. (London)* 44, 275
7 Goodford, P. J. (1977) in *Drug Action at the Molecular Level* (Roberts, G. C. K. ed.), pp. 109–126, Macmillan, London
8 Goodford, P. J., St-Louis, J. and Wootton, R. (1978) *J. Physiol. (London)* 283, 397–407
9 Greenwald, I. (1925) *J. Biol. Chem.* 63, 339
10 Perutz, M. F., Muirhead, H., Cox, J. M. and Goaman, L. C. G. (1968) *Nature (London)* 219, 131–139

Receptor-secretion coupling in mast cells

John Foreman

Department of Pharmacology, University College London, Gower Street, London WC1E 6BT, U.K.

The cross-linking of membrane receptors in mast cells and basophil leucocytes initiates a series of biochemical changes, involving methylations in membrane lipids. The result of these events is the increase of membrane permeability to calcium which leads to the secretion of histamine.

The mast cell is a secretory cell which liberates its histamine-containing granules in response to a number of specific stimuli. It has proved to be a powerful and interesting model in the study of secretion, and the release of histamine from mast cells following an immunological stimulus to the membrane is a basic pathological mechanism of allergic disease.

The membrane stimulus

Mast cells and their circulating analogues, basophil leucocytes, possess membrane receptors which bind specifically the so-called homocytotropic antibody, now known to be IgE in most species[1]. There are about 1 to 5×10^5 IgE receptors per cell and the equilibrium dissociation constant for the binding of IgE to its cellular receptor is about 10^{-11} M. The forward rate constant of the binding reaction is estimated to be of the order of 10^4 M^{-1} s^{-1} and the reverse reaction is relatively slow with a rate constant of 10^{-7} s^{-1}.

The binding of IgE antibody to its cellular receptor does not in itself bring about activation of the cell. Indeed, mast cells have varying degrees of saturation of their IgE receptors, depending on the circulating IgE level. The signal to the mast cell or basophil leucocyte to secrete histamine-containing granules is the union of the cell-bound IgE antibody with antigen present in the surrounding milieu. The antigen must, of course, be that to which the IgE has specifically-directed binding sites (Fig. 1). It has long been appreciated that monovalent antigens can combine with only one of the two antigen binding sites that each IgE molecule possesses (Fig. 1a) and this does not bring about activation of the cell for secretion. To be active, an antigen must be divalent or multivalent, an observation which can be interpreted in terms of bridging of two antigen binding sites either within a single IgE molecule (Fig. 1b) or between adjacent IgE molecules (Fig. 1c).

Other evidence suggests that it is indeed the bridging of adjacent IgE molecules which activates the secretory process of the mast cell or basophil leucocyte. The evidence may be summarized as follows:

(1) Antibody (anti-IgE) directed against the F_c portion of the IgE molecule cross-links adjacent IgE molecules and induces histamine secretion (Fig. 1d).

(2) Concanavalin A binds to the carbohydrate associated with the F_c region of the IgE molecule, thereby cross-linking adjacent molecules and inducing histamine secretion (Fig. 1e).

(3) Chemically dimerised IgE molecules added to the extracellular fluid bind to unoccupied IgE receptors on the

cell surface, inducing cross-linking of IgE receptors, and histamine secretion occurs (Fig. 1f).

(4) The IgE receptor, obtained from a rat basophil leukaemia, has been isolated and purified. It is a glycoprotein with a molecular weight of about 80,000 and an antibody has been prepared against it. The antireceptor antibody (an IgG) is divalent and can cross-link adjacent receptors for IgE on the cell membrane. Antireceptor antibody stimulates histamine secretion (Fig. 1g).

Whilst the evidence for the cross-linking of IgE receptors in the cell membrane being the stimulus to secretion is compelling, an alternative hypothesis should be mentioned. It is based on the observation that certain peptides derived from the IgE antibody molecule are capable of eliciting histamine secretion, and the hypothesis states that binding of antibody to the receptor does not itself activate secretion, but conformational change in the antibody brought about, say, by antigen binding, can expose these peptide fragments which

induce release. Although the hypothesis does not easily explain the results of experiments such as those in Fig. 1a and 1g, there is no solid evidence against it.

The IgE receptors, 100 to 500 × 10[3] of them per mast cell appear to be monovalent with respect to IgE binding, but, so far, no biochemical activity has been associated with the receptor, even in its purified form. It is a glycoprotein of molecular weight about 80,000 and there is some dispute about whether it consists of two subunits. Some elegant experiments employing photobleaching have demonstrated that the receptor is freely mobile in the lateral plane of the membrane[1].

Calcium

It has been known for some time that histamine secretion from mast cells stimulated by the cross-linking of antibody on the membrane requires the presence of extracellular calcium, and it was suggested by analogy with muscle that a rise in intracellular calcium ion concentration might be the link between membrane

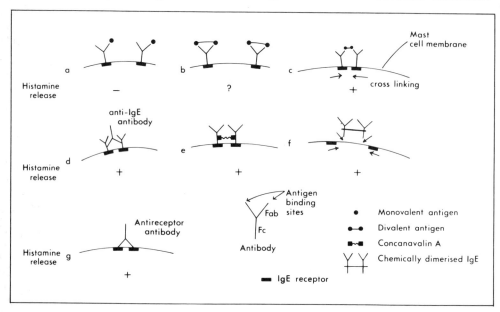

Fig. 1. Diagrammatic representation of antigen-IgE binding on the mast cell membrane and other mechanisms of cross-linking IgE receptors in the cell membrane. Those reactions leading to cell activation and secretion are indicated by +.

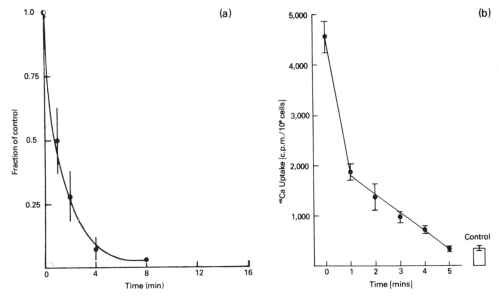

Fig. 2. (a) Time-course of the decay of mast cell response to calcium following an antigen-IgE stimulus. Histamine secretion is measured as a function of increasing interval between stimulation in the absence of calcium and the addition of calcium. (b) Time-course of the change of ^{45}Ca uptake after antigen stimulation. Cells from the same pool were either challenged with antigen in the presence of ^{45}Ca: t = 0, or cells were challenged with antigen in the presence of non-labelled calcium and ^{45}Ca was added at t min after stimulation. The control represents ^{45}Ca uptake in the absence of stimulation.

stimulus and histamine secretion. The principal evidence for this hypothesis came from experiments with the calcium ionophore A23187[2] which transports calcium across the mast cell membrane from extracellular to intracellular compartments and thereby induces histamine secretion. Experiments with the ionophore demonstrated a major role for extracellular calcium ions in the initiation of histamine secretion, although it has been suspected for some time that release of calcium from intracellular stores may induce secretion in certain circumstances, and recent experiments with A23187 are in keeping with this[3].

Even before the ionophore was available it had been suggested that the immunological reaction on the mast cell membrane opened channels to allow calcium to pass from the extracellular compartment to the intracellular compartment. Using radiolabelled calcium, it has been shown that the resting permeability of the mast cell membrane to calcium is about 8 fmol cm^{-2} s^{-1} and that this rises to a peak of 380 fmol cm^{-2} s^{-1} after an optimal immunological stimulus. It has recently been demonstrated that direct cross-linking of IgE receptors themselves, using an antireceptor antibody, is sufficient to increase the membrane permeability to calcium and thereby induce histamine secretion[4].

There has been no measurement of the rise in free cytosolic calcium which is presumed to initiate histamine secretion, but it is known that injection of calcium into mast cells using a micropipette will induce the exocytosis of granules[5]. A further piece of evidence consistent with the hypothesis that a rise in intracellular calcium concentration triggers histamine secretion comes from experiments employing phospholipid vesicles loaded with calcium. Fusion of these vesicles with mast cells induces histamine secretion presumably by liberating calcium into the intracellular compartment[6].

Inactivation

It is reasonable to expect that any cell employing a rise of intracellular calcium concentration for activation will also possess a mechanism for sequestering calcium, so that the activity does not continue indefinitely. Mast cells appear to possess a mechanism for turning off the increase in membrane permeability to calcium which the immunological stimulus brings about. Fig. 2 shows that after immunological stimulation of the membrane, the membrane permeability to calcium and the secretion of histamine decline with a half-time of about 1 min. Since secretion is largely complete 30 s after the stimulus, the decay of membrane calcium permeability cannot be the only means of terminating or limiting secretion and it is assumed that the action of calcium is rapidly terminated by pumping into mitochondria and other organelles or out of the cell. It is worth noting that the inactivation of membrane permeability to calcium is not the result of dissociation of antigen from IgE or of IgE from its cellular receptor. Furthermore, there is no apparent redistribution, 'capping' or endocytosis of receptors during the time when inactivation is occurring. Some recent experiments show that the entry of calcium into the mast cell induces specific phosphorylation of a membrane protein, and it is interesting to speculate that this may be related to the formation and inactivation of calcium channels.

Phospholipids

A role for phospholipid in histamine secretion was identified before the current awareness of the phenomena of membrane activation such as phosphatidyl inositol turnover, and before the interpretation of receptor function in terms of the fluid-mosaic model for membranes. Immunologically-triggered histamine secretion from mast cells has been shown to be potentiated specifically by phosphatidyl serine or its lyso-analogue; the lipid itself not being capable of stimulating secretion. Furthermore, this action of phosphatidyl serine was clearly related to the role of calcium in histamine secretion and it has been shown that phosphatidyl serine increases the membrane permeability to calcium in cells stimulated by a cross-linking stimulus. The potentiating effect of phosphatidyl serine in immunologically mediated secretion is associated with the binding of vesicles of the phospholipid to the mast cells. It is interesting that a number of N-substituted derivatives of phosphatidyl serine, which are not metabolized, are not active on mast cells[7].

These observations fit very well into a scheme proposed by Axelrod and colleagues (see Fig. 3). Cross-linking of IgE receptors on the mast cell initiates a sequence of membrane phospholipid metabolism and enzyme activation which results in the formation of calcium channels. The initial event is decarboxylation of phosphatidyl serine to form phosphatidyl ethanolamine which is then methylated by a magnesium-dependent methyltransferase I to produce an N-monomethyl derivative. A second methyltransferase converts the methyl derivative of phosphatidyl ethanolamine into phosphatidyl choline which serves as a substrate for a membrane phospholipase A_2. The lysophosphatidyl choline produced is considered to be responsible for the opening of a calcium channel. High concentrations of lysophosphatidyl choline release histamine from mast cells by a mechanism independent of calcium and in which the cells are lysed, but this effect on calcium channels is considered to be a more selective and specific effect. The methylation reactions have been measured in mast cells and increased methylation occurs in immunologically stimulated cells. Phospholipid methylation precedes calcium exchange across the membrane and is independent of the presence of extracellular calcium[8]. Inhibitors of the methyltransferases such as S-iso-butyryl-3-deaza-adenosine prevent both calcium exchange across the membrane and histamine secretion. Thus, the original observation that phosphatidyl serine potentiates histamine secretion and increases membrane permeability to calcium following receptor cross-linking, may now be interpreted in terms of phosphatidyl serine being a limiting substrate in the phospholipid metabolism necessary for calcium channel formation.

Phosphatidyl inositol turnover in mast cells increases when the cells are stimulated by cross-linking of receptors but the increased turnover is both dependent on calcium and stimulated by phosphatidyl serine, hence there is no evidence that phosphatidyl inositol turnover is anything more than an epiphenomenon of mast cell activation[9].

Cyclic AMP

It has been proposed that inactivation of calcium channels may limit, in part, secretory activity in mast cells. The question of the mechanism of this inactivation therefore arises. Cyclic AMP inhibits histamine secretion, and there is evidence which suggests that a fall in intracellular cyclic AMP accompanies the activation of secretion. As the levels of cyclic AMP return towards their basal value, inactivation of calcium channels is occurring. It has been demonstrated that cyclic AMP prevents the immunologically stimulated calcium exchange across the mast cell membrane and it is, therefore, tempting to suggest that a fall in intracellular cyclic AMP level occurs during calcium channel formation and the induction of histamine secretion, whilst a return to basal level of this cyclic nucleotide inactivates calcium channels

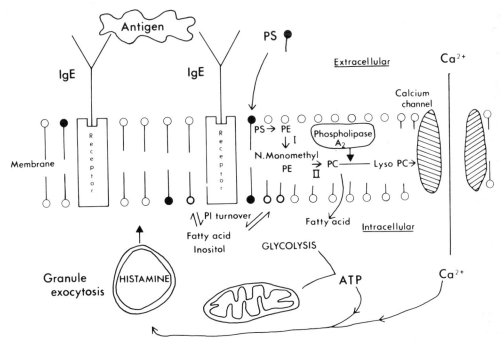

Fig. 3. A diagrammatic summary of the membrane events considered to be involved in the activation of mast cells to secrete their histamine. Receptor cross-linking initiates a sequence of phospholipid methylations and the activation of a phospholipase which together bring about the opening of calcium channels. Cyclic AMP (not shown) is believed to be responsible for inactivating the channel, possibly by a protein kinase mediated phosphorylation. (I and II indicate the two methyltransferase enzymes.)

and limits secretion. Mast cells possess a protein kinase which is activated by cyclic AMP and may be responsible for the phosphorylation of membrane protein (see above) during inactivation of calcium channels.

The events discussed in this review relate to very early stages in the secretion of histamine and its control (Fig. 3). It would be naive to pretend that they are more than a fraction of the whole picture. Little is known about the events occurring after the entry of calcium other than that they require a source of ATP. It seems fairly clear that microtubules are not involved in the mechanism of histamine secretion[10], but actin has been demonstrated in mast cells and indirect evidence for a role for microfilament contraction in secretion has been presented.

Reading list

1 Metzger, H. (1977) in *Receptors and Recognition* series A, Vol. 4 (Cuatrecasas, P. and Greaves, M. F., eds), pp. 75–102, Chapman Hall, London

2 Foreman, J. C., Mongar, J. L. and Gomperts, B. D. (1973) *Nature (London)* 254, 249–251

3 Johansen, T. (1980) *Eur. J. Pharmacol.* 62, 329–334

4 Ishizaka, T., Foreman, J. C., Sterk, A. R. and Ishizaka, K. (1979) *Proc. Natl. Acad. Sci. U.S.A.* 76, 5858–5862

5 Kanno, T., Cochrane, D. E. and Douglas, W. W. (1973) *Can. J. Physiol. Pharmacol.* 51, 1001–1004

6 Theoharides, T. C. and Douglas, W. W. (1978) *Science* 202, 1143–1145

7 Martin, T. W. and Lagunoff, D. (1979) *Science* 204, 631–633

8 Ishizaka, T., Hirata, F., Ishizaka, K. and Axelrod, J. (1980) *Proc. Natl. Acad. Sci. U.S.A.* 77, 1903–1906

9 Cockroft, S. and Gomperts, B. D. (1970) *Biochem. J.* 178, 691–687

10 Lagunoff, D. and Chi, E. Y. (1976) *J. Cell Biol.* 71, 182–195

Sweeteners and receptor sites

Guy A. Crosby and Grant E. DuBois

Chemical Synthesis Laboratories, Dynapol, Palo Alto, CA 94304, U.S.A.

Since the turn of the century, scientists have been both fascinated and confounded by the vast array of chemical structures that are capable of eliciting a sweet taste response in humans[1]. A small sample of more than 30 types of molecules reported to be sweet to the taste is illustrated in Fig. 1. Note the striking variation in size, shape and functionality between these compounds. This seemingly insensitive relationship between structure and taste stands in stark contrast to the observation that even minor chemical modifications of sweet-tasting compounds frequently result in a significant change or loss of taste[2]. Consider, for example, the analogues of saccharin, (+)-tryptophan and aspartame shown in Fig. 1. Saccharin can be converted into 6-hydroxysaccharin without loss of sweetness, yet 6-methoxysaccharin is tasteless[1]. Replacement of the amide nitrogen in aspartame with an oxygen results in loss of sweet taste[3]. Exchange of chlorine for hydrogen at the 6-position of (+)-tryptophan increases sweetness nearly 40 times[4]. How can we resolve this paradoxical relationship between taste and molecular structure in light of the current views on the mechanism of taste?

Structure–taste relationships based on receptor theory

A major advance in the interpretation of taste–structure relationships came with the understanding that the primary event in the initiation of a taste response involves the interaction of a stimulant molecule with a receptor located at the taste cell plasma membrane. As early as 1919, Renqvist proposed that the initial step takes place on the surface of the taste cell membrane[5]. More recent evidence is also in favour of the taste cell stimulation being extracellular in nature, with the stimulant molecule–receptor interaction occurring on the microvillus membrane of taste cells (Fig. 2)[1].

Beidler proposed in 1967 that sweet taste stimuli reversibly interact with receptor cells by hydrogen bonding to proteins of the cell membranes[6]. It was hypothesized that the hydrogen bonding caused local reorientation of the cell membrane, creating channels through which ions could pass, thus depolarizing the receptor cells. In the same year (1967), Shallenberger and Acree proposed the first correlation of structural features with sweet taste[7]. Their model stated that all sweet compounds contained an A-H/B binding unit, where A and B are electronegative atoms separated by 2.4–4.0 Å, and H is a hydrogen atom. Furthermore, this A-H/B system was thought to be involved in forming simultaneous hydrogen bonds to a complementary hydrogen donor–acceptor system on the receptor. The Shallenberger model for sweetness has been elaborated upon by Kier for intense sweeteners[8]. Kier believes that basal levels of sweetness may be obtained with a simple A-H/B unit, but that intense sweetness cannot be obtained without a third binding region, which may take the form of a hydrophobic group. The

Glycine
MW 75

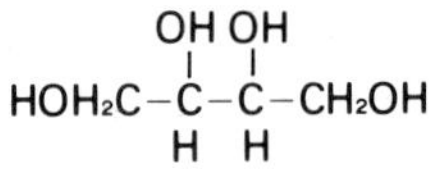

Erythritol
MW 122

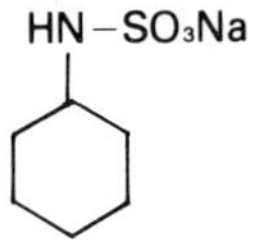

Cyclamate
MW free acid 179

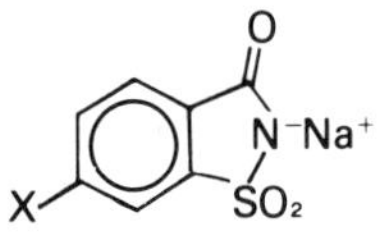

Saccharin (X = H)
MW free acid 183
X = OH, sweet
X = OCH₃, tasteless

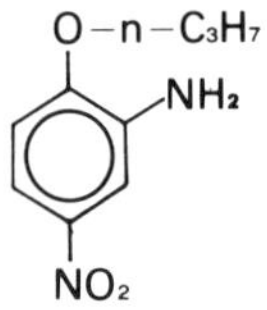

P-4000
MW 196

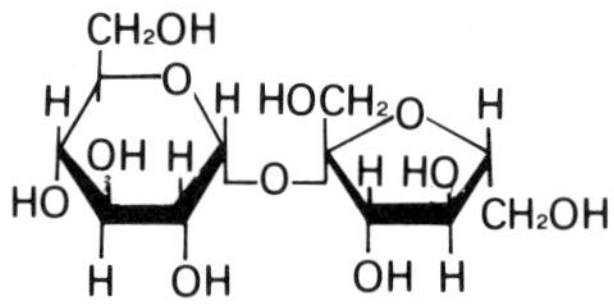

Sucrose
MW 342

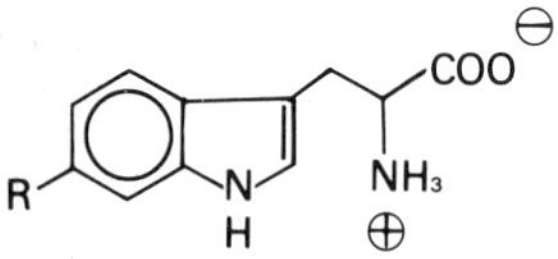

(+)-Tryptophan (R = H)
MW 204
(R = Cl is 37× more
potent than R = H)

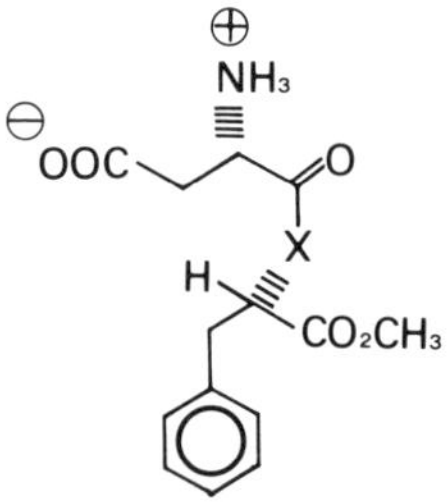

Aspartame (X = NH)
MW 280
(X = O, tasteless)

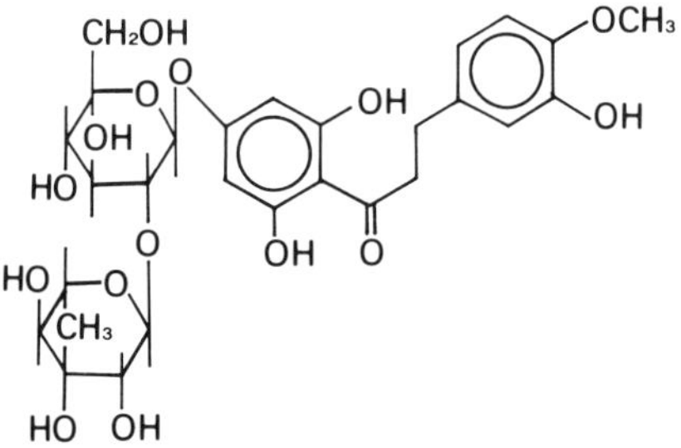

Neohesperidin dihydrochalcone
MW 612

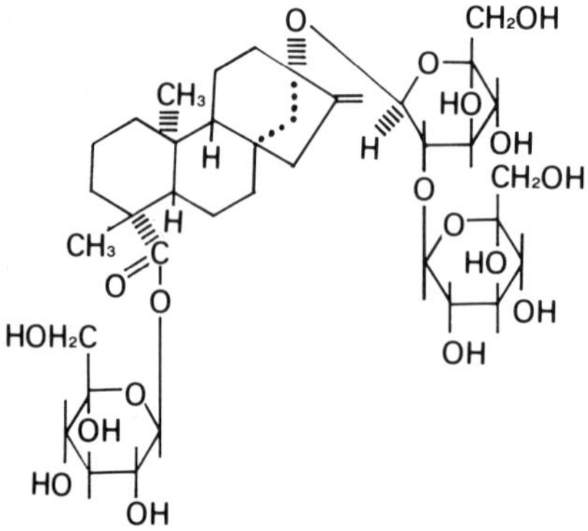

Stevioside
MW 805

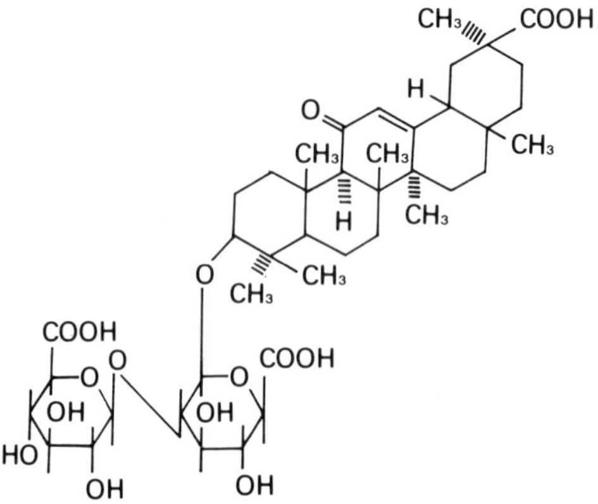

Glycyrrhizin
MW 823

Proteins: monellin MW 10 700; thaumatin I and II MW 22 000.

Fig. 1. Examples of natural and synthetic sweeteners. MW = molecular weight.

preferred conformations of sweet D-amino acids, such as tryptophan, phenylalanine, histidine and leucine, were calculated by molecular orbital methods, and all were found to contain regions of high electron density in the same position relative to the zwitterion moiety (Fig. 3a). These positions were therefore implicated as third (X) binding points and suggested modification of the A-H/B unit, or essential glucophore for sweet taste, to that illustrated in Fig. 3b.

Is there more than one receptor site for sweet taste?

Unfortunately, the idea that all sweeteners contain a common essential glucophore has been interpreted to mean there is a common receptor site for all sweeteners. The validity of this assumption is very uncertain at this time for the human gustatory system. It seems more likely that sweet substances having structures as diverse as those in Fig. 1 are exerting their effects at different receptor sites in humans. For insects, the presence of different receptors for different sugars has been implicated. Studies with the *Drosophila* fly[9] and fleshfly[10] have led to the proposal of two receptor sites for monosaccharides, with one sensitive to D-glucose and the other sensitive to D-fructose and various disaccharides. In humans, there are some sweeteners, like sucrose and the amino acids, that elicit a response mainly on the forward part of the tongue, and others, like neohesperidin dihydrochalcone, that elicit a sensation on the tongue as well as the back of the mouth. Boudreau[11] has recently classified at least two types of sweet taste sensations, identified as sweet1 and sweet2, depending on the location of the sensation. If two sweet taste receptors, call them S_1 and S_2, do exist in different parts of the oral cavity, it may be of some value to consider the potential analogy of these receptors with the α- and β-adrenergic receptors and the H_1- and H_2-histamine receptors.

To make sense out of this confusing picture of so many diverse structures, all producing the same biological effect, it seems reasonable to assume that one or more receptor macromolecules, presumably proteins, may provide an array of structurally related sites that respond to a variety of sweet substances, rather than a common receptor site. Different binding sites for various classes of sweeteners would be found on different regions of the receptor macromolecules. Thus, the nature of the receptor site would remain relatively constant in character within one class of sweeteners. In other words, one receptor site may exist for sugars, another for amino acids, and yet another for dihydrochalcones. The recent efforts of Kier to determine the nature of the receptor feature responsible for binding with the nitroaniline class of sweeteners can be viewed in this light. Kier envisions the 2-substituent of 5-nitroanilines (P-4000, Fig. 1 is an example of a 2-substituted 5-nitroaniline) as being involved in dispersion binding (van der Waals interaction) with some component of the receptor[12]. Interaction energies were calculated as a sum of electrostatic (E_c), polarization (E_p),

TABLE I. Sweetness–structure relationships among 2-substituted 5-nitroanilines and total calculated interaction energies (reprinted with permission from Höltje and Kier[12]).

Substituent position −2	Long sweetness relative to sucrose	Calculated total binding energy at 4.25 Å distance (kcal mol⁻¹)
OC₃H₇	3.61	5.40
OCH₂CH = CH₂	3.30	5.25
I	3.10	4.93
OCH₂CH₃	2.98	4.91
Br	2.90	3.79
OCH(CH₃)₂	2.78	4.91
Cl	2.60	3.59
OCH₃	2.34	4.22
CH₃	2.34	3.96
F	1.50	3.24
H	1.50	3.11

dispersion (E_d), and repulsion (E_r) energies for a series of cogeners and several models for protein side chains. For example, 3-methylindole was used as a model for the tryptophan side chain of a protein receptor.

A correlation was then sought between the interaction energies, at a constant substituent-receptor site distance, through the series and the sweetness level. Only the model for the tryptophan side chain

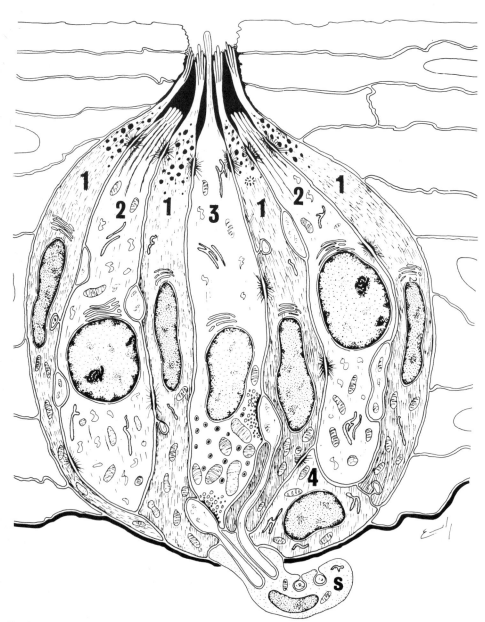

Fig. 2. A longitudinal section of a taste bud from a rabbit foliate papilla. Surface epithelial cells (outline only) are penetrated by the outer taste pore. The finger-like projections extending into the outer taste pore are the microvilli (0.2 × 2.0 μm) of the taste cells. The numbers indicate the different types of taste cells (light and dark). Reproduced with permission from Murray, R. G. (1973) in The Ultrastructure of Sensory Organs (Friedman, I., ed.), pp. 1–8, Elsevier/North-Holland Biomedical Press, Amsterdam.

yielded a significant correlation. The mode of substituent–receptor approach and distance were then optimized to give the best correlation between total interaction energy and sweetness, as shown in Table I. It is reasonable to assume that other classes of sweeteners could be correlated with other protein side chain residues, based on the size, shape and functionality of the sweet stimulus.

As a result of Kier's work, it appears that a tryptophan residue may be an important component of the receptor site for-nitro-aniline sweeteners where the side chain of the amino acid is involved in a dispersion binding interaction with the 2-substituent of the nitroaniline system. Consideration of additional hydrogen donor–acceptor interactions of the type proposed by Shallenberger suggests a hypothetical binding interaction between 2-n-propoxy-5-nitroaniline and the nitroaniline receptor, as shown in Fig. 4.

A model for the sweet taste receptor

Kier's work sets the stage for developing an attractive model for the sweet taste receptor, which can accommodate the vast array of sweet stimuli, based on the early concept of Fischer[13]. He envisions the process of receptor stimulation being similar to the reversible changes of Gram-positive to Gram-negative staining behaviour, which occurs in wool and bacterial membranes. Fischer proposes that the initial steps in taste sensation might be governed by the affinity of a stimulus molecule for an α-helical region of a protein, which then undergoes transition to a β-pleated sheet conformation. Desorption of the chemical stimulus by saliva allows relaxation to the original α-helical receptor. Now it is well known that intramolecular hydrogen bonding is important in the formation of the α-helical resting state of a protein, while proper interaction between amino acid side chains is necessary for stabilization of this ordered conformation[14]. Thus, proteins containing hydrophobic side chains are stabilized in the α-helical state, whereas charged proteins, such as polylysine and polyglutamic acid, exist predominantly in a random coil owing to charge repulsions of the side chains. From Kier's work, it is reasonable to speculate that sweet stimulant molecules may be involved in breaking up the stabilizing side chain interactions present in the α-helix and inducing a trans-

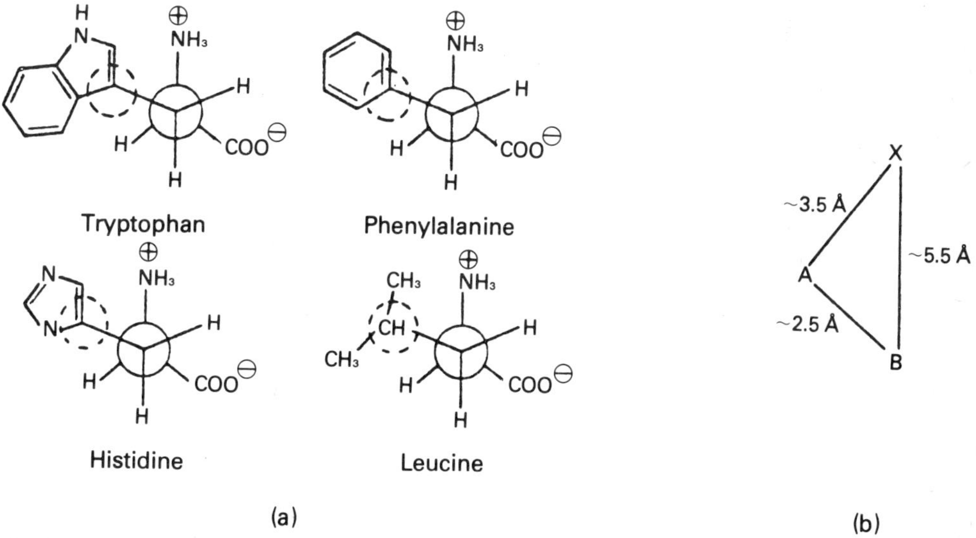

Fig. 3. (a) D-Amino acid side chain features found to occupy a common spatial area relative to the zwitterion moiety; (b) pattern of atoms imparting a sweet taste (glucophore) postulated from this study.

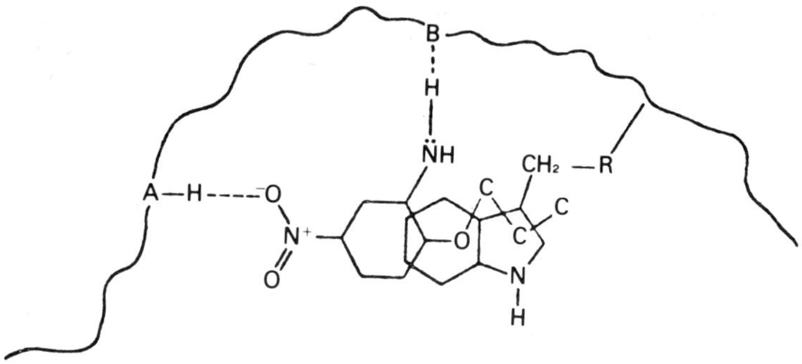

Fig. 4. Hypothetical binding interactions between 2-n-propoxy-5-nitroaniline and the m-nitroaniline receptor. Reproduced with permission from Crosby *et al*[1].

ition to β-pleated sheet or random coil states, leading to a depolarization of the receptor cell as suggested by Beidler[6]. It is encouraging to note that tryptophan, identified as a potential binding site for the nitroaniline sweeteners, is known to stabilize the α-helical conformation of proteins[14]. Other classes of sweeteners may bring about an identical conformational change of the same receptor protein through the interaction with other side chain substituents occurring in different regions of the helix, all culminating in the response of sweet taste. Such a helix therefore has many potential binding or receptor sites. This model is attractive in that it rationalizes both the diversity and specificity of structure occurring among sweet compounds. It also illustrates the role of multiple receptor sites which may yield a common biological response when bound to different classes of sweet substances.

Many features of this model are similar to the concept of a generalized receptor site advocated by Beets[15]. We have developed a more detailed picture of the early stages of receptor–substrate interaction as it relates to a common mechanism for the many types of sweet substances[1]. In common with Beets, we believe the macromolecules serving as receptors on the membrane of receptor cells may be present on the membranes of other cells. Attempts to isolate a 'sweet sensitive protein' suggest

this reasoning. However, the particular macromolecule can serve as a sweet taste receptor only if the cell on which it occurs functions as a receptor cell. When it occurs on the membrane of any other type of cell, it may act with molecules in the same way, but such interaction has no function because the cell is not equipped to initiate a taste response.

Reading list

1 Crosby, G. A., DuBois, G. E. and Wingard, R. E. (1979) in *Drug Design* (Ariëns, E. J., ed.), Vol. 8, pp. 215–310, Academic Press, New York.

2 Crosby, G. A. (1976) *Crit. Rev. Food Sci. Nutr.* 7, 297–323

3 MacDonald, S. A., Willson, C. G., Chorew, M., Vernacchia, F. S. and Goodman, G. (1980) *J. Med. Chem.* 23, 413–420

4 Kornfeld, E. C., Suarez, T., Edie, R., Brannon, D. R., Fukuda, D., Sheneman, J., Todd, G. C. and Secondino, M. (1974) *Abstr. of Papers, Amer. Chem. Soc.* 16, 41

5 Renqvist, Y. (1919) *Skand. Arch. Physiol.* 38, 97–201

6 Beidler, L. M. (1967) in *Olfaction and Taste II* (Hayashi, T., ed.), pp. 509–534, Pergamon, New York

7 Shallenberger, R. S. and Acree, T. E. (1967) *Nature (London)* 216, 480–482

8 Kier, L. B. (1972) *J. Pharm. Sci.* 61, 1394–1397

9 Isono, K. and Kikuchi, T. (1974) *Nature (London)* 248, 243–244

10 Shimada, I., Shiraishi, A., Kijima, H. and Morita, H. (1974) *J. Insect Physiol.* 20, 605–621

11 Boudreau, J. C., Oravec, J., Hoang, N. K. and White, T. D. (1979) in *Food Taste Chemistry*

(Boudreau, J. C., ed.), pp. 1–32, American Chemical Society, Washington, D.C.

12 Höltje, H.-D. and Kier, L. B. (1974) *J. Pharm. Sci.* 63, 1722–1725

13 Fischer, R. (1971) in *Gustation and Olfaction* (Ohloff, G. and Thomas, A. F., eds), pp. 187–237, Academic Press, New York

14 Lehninger, A. L. (1970) *Biochemistry*, pp. 113–114, Worth, New York

15. Beets, M. G. J. (1978) *Structure–Activity Relationships in Human Chemoreception*, pp. 35–41, Applied Science, London

Index

Acetycholine receptor   34
Acetylcholine regulator   34
Adenylate cyclase
  and β-adrenoceptor   53, 61
  and antidepressants   99
  and desensitization   28
  dopamine sensitive   105, 112
  inhibition   90
  and phospholipid metabolism   132
Adrenergic mediator, history   44
Adrenoceptors
  and antidepressants   99
  classification   49
α-Adrenoceptors   84
β-Adrenoceptors
  characterization   53
  regulation   61
  subtypes   61, 71, 78
Adrenocorticotropin   139
γ-Aminobutyric acid receptors   171, 176, 184
Androgen binding   192
Angiotensinase   147
Angiotensin receptor   147
Antidepressant drugs   99
Autoinhibition, α-adrenergic   94
Autoregulation
  of vitamin D system   200

Benzodiazepines, action   176
Benzodiazepine receptors   176, 184
Beta blockers   49, 78
Beta stimulants   78
Binding energy
  of drugs   8
Binding reactions,
  rate of   16
Binding sites
  and allosterism   34
  opiate receptor   159
Binding studies
  and α-adrenoceptors   84
  and β-adrenoceptors   61, 71, 78
  and dopamine receptors   105
  and H_1 histamine receptor   166
  steroid hormones   192

Calcium
  in histamine secretion   217

Calcium binding protein   200
Calcium homeostasis   200
Calcium signalling   122,132
Cannon, W. B.   44
β-Carbolines   184
Cardiovascular shock syndromes
  and dopamine   118
Central nervous system
  adrenoceptors   99
  benzodiazepine receptors   184
  dopamine receptors   105, 112, 118
  GABA receptors   171, 176
  histamine H_1 receptors   166
  opiate receptors   159
Chemoreceptor theory   1
Classification
  adrenoceptors   44, 49
  α-adrenoceptors   84
  β-adrenoceptors   61, 71, 78
  dopamine receptors   105, 118
  GABA receptors   171, 176
  opiate receptors   151, 159
Concentration jump relaxation   16
Cyclic AMP
  and desensitization   28
  and dopamine receptors   112
  and histamine secretion   217

Desensitization   28
Diazepam binding   184
Dihydroalprenolol binding   71, 78
Dihydroergocryptine binding   84
Dihydroxyvitamin D_3 receptors   200
Dopamine receptor
  characterization   105
  classification   105, 118
  and cyclic AMP   112
Drug action
  antidepressants   99
  benzodiazepines   176
  rate of   16
Drug design   205
Dyes and receptor theory   1
Dyskinesia   118

Ehrlich, Paul   1
Electroplaque synapse   34
Endocrine disorders
  and dopamine   118

Endocrine system
  vitamin D   200
Endorphins   151, 159
Enkephalins   151, 159
Estrogen binding   192

GABA-modulin   176
Glucocorticoid binding   192
Glucophores   223
Guanine nucleotide
  binding protein   53, 90, 122
  regulation   61, 84, 90, 112
Guanylate cyclase   122

Haemoglobin as receptor model   205
Haloperidol binding   105
Histamine H₁ receptors   166
Histamine secretion   217
Hormone receptors
  and adenylate cyclase   90
  excitation   8
5-Hydroxytryptamine
  and antidepressants   99

Immune response
  and desensitization   28
Immunoglobulin E receptors   217

Langley, John Newport   1
Lipid bilayers
  and adrenocorticotropin   139

Mast cells   217
α-Melanocyte stimulating hormone   112
Mepyramine binding   166

Neuroleptic drugs   105
Neuromodulators   94, 176, 184
Neuromuscular junction   34
Nicotine-receptive substance   1
Nicotinic receptor
  characterization   34
  desensitization   28
Noise analysis   16
Noradrenaline
  and antidepressants   99
  release   94
  and sympathin E   44
Noradrenergic neurones   94

Opiate receptors
  classification   151, 159
  in CNS   159
  peripheral   151
Opioid peptides   151
Oxygen affinity states   205

Parathyroid hormone   200
Parkinson's disease   118
Peptide information, transduction   139
Phosphate homeostasis   200
Phosphatidylinositol hydrolysis   122, 132
Phospholipid metabolism   122, 132
  in mast cells   217
Plasma membrane
  and angiotensinase   147
  and hormone receptors   90
Presynaptic receptors, noradrenergic   94
Progestin binding   192
Purification
  acetylcholine receptor   34
  β-adrenoceptor   53, 61
  immunoglobulin E receptor   217

Receptors
  acetylcholine   34
  adrenergic   49
  α-adrenergic   84
  β-adrenergic   53, 61, 71, 78
  adrenocorticotropin   139
  γ-aminobutyric acid   171, 176, 184
  angiotensin   147
  benzodiazepine   176, 184
  calcium-mobilizing   122, 132
  and calcium signalling   122
  and desensitization   28
  dihydroxy vitamin D   200
  dopamine   105, 112, 118
  histamine H₁   166
  hormone   8, 90
  immunoglobulin E   217
  opiate   151, 159
  peptide hormone   139
  presynaptic   94
  steroid hormone   192
  sweet   223
Receptor excitation   8
Receptor model
  haemoglobin as   205
  sweet taste receptor   223
Receptor proteins   34
Receptor-secretion coupling   217
Receptor theory, origins   1
Reconstitution
  of acetylcholine regulator   34
Renin-angiotensin system   147
Rosenblueth, A.   44

Schizophrenia
  and dopamine   118
Secretion, mast cells   217
Side chain theory   1
Single ion channels   16

Solubilization
  α_1-adrenoceptors 84
  β-adrenoceptors 53, 61
  dopamine receptors 105
  opiate receptors 159
Steroid hormone receptors 192
Substance P 159
Sweeteners 223
Sympathins, history 44

Taste receptors 223
Three-state protein model 205
Toxin-antitoxin interaction 1

Vasoactive peptides 151
Vitamin D endocrine system 200
Voltage jump relaxation 16

Towards Understanding Receptors

A collection of readable reviews on one of the most important topics in pharmacology and biochemistry of this decade. Progress towards elucidation of the events occurring between the interaction of drugs with receptors and their effects has been dramatic in recent years and this volume records many of the key discoveries in the field. In his foreword G. Alan Robison gives an overview which lends coherence to the whole field of receptor research. This volume provides essential and affordable reading for students, pharmacologists, biochemists, physicians and all others who wish to keep abreast of this vital area.

£7.00   ISBN 0-444-80339-4